1. 330-ton Crane #15

2. 220-ton Crane #11

3. A and B Ways

4. Aluminum Shop

5. Assembly Building

6. Main Gate

7. West Gate

8. South Gate

9. Unit Storage

10. Preoutfit-II (PO-2)

11. Blast and Paint Facility

12. North Offices: Engineering/Drafting/ Lofting/Planning

13. Pipefitters

14. Main Administration Building

15. Machinists

16. Electricians

17. South Pier

18. Outfitting Pier

19. North Pier

20. North Yard Assembly

21. AEGIS Command SUPSHIP Offices

22. Precommissioning Unit Offices

THE YARD

THE YARD

Building a Destroyer at the Bath Iron Works

MICHAEL S. SANDERS

HarperCollins*Publishers*

Photograph and Illustration Credits

Figure 2 by Kevin Callahan.
Figures 3, 22, 36, and endpaper aerial photo
    ©1998 Bath Iron Works/A General Dynamics Company.
Figures 5, 6, and 7 based on drawings used with permission of Bath Iron
    Works Corporation, adapted by Kevin Callahan.
Figures 24, 25, 26 ©1998 John Davis Bidwell.
Figure 37 ©1998 Lockheed Martin.
All other photographs © 1998 by J. Atherton Monroe.

FIRST EDITION

Designed by Nancy B. Field

Library of Congress Cataloging-in-Publication Data

Sanders, Michael S.
    The yard : building a destroyer at the Bath Iron Works / Michael S. Sanders.
— 1st ed.
        p.   cm.
    Includes index.
    ISBN 0-06-019246-1
    1. Destroyers (Warships) — United States. 2. Bath Iron Works. I. Title
V825.3.S27 1999
623.8'25'0973—dc21   99-15026

00 01 02 03 ❖/RRD 10 9 8 7 6 5

For my wife, Amy Beth Russell.

Always remember . . .
Là, tout n'est qu'ordre et beauté, luxe, calme, et volupté.
And someday we'll find it.

CONTENTS

INTRODUCTION

From the center span of the Carlton Bridge over the Kennebec River in Bath, Maine, it's about a thousand yards to the sprawling complex of building ways, cranes, and assembly hangars that constitute the Bath Iron Works shipyard on the western bank. At three o'clock in the morning the night before the yard launches a ship, with a clear sky and a full moon above, it seems like fifty feet. There is no wind then, and the water, smooth as glass, mirrors in pinpoint clarity every light blazing in the shipyard, the dirty pink of the sodium vapors, the blinking reds atop the cranes, the periodic white flashes of the strobes mounted on the yardarms of the Navy destroyers tied up at the pierways jutting out in a shallow V into the river downstream.

Viewed this way, obliquely from one end, the yard resembles a child's Erector set, a toy city left out in a jumble on the playroom floor at bedtime, all odd angles and thrusting fingers of steel, on the ground barely discernible everyday objects—a stake-bed truck, 55-gallon drums, a pile of steel pipe, wooden beams lying in a tangled heap. Closer in, there is an order to this chaos. BIW runs from the foot of the bridge down the waterfront about a mile and a half out towards the edge of town, but penetrates back from the shore only narrowly, with the two-story houses and storefronts of Washington Street butting right up against its back.

As the eye travels down the waterfront from the bridge, it first meets the red-trimmed tugs snugged up to the dock by the machine shop, then the white iron tracery of Crane #15 soaring like an upraised arm hundreds of feet in the air. Beyond lies the first of three elevated, steel-reinforced concrete ramps that run at an angle five hundred and thirty feet straight down into the river—the "ways" on which the ships are constructed.

Crane #11, with an even more massive base, rises to an even greater height between the second and third ways, by which time the eye has become lost in the obscuring confusion of scaffold staging, cables, wooden and metal supports, and heavy-duty nylon drapes hung seemingly at random from a very large object whose briefest outline is defined by strings of incandescent bulbs strung in a line thirty feet above the concrete pavement. This night, the night of a launch, the structure crawls with the shadowy shapes of men in hardhats busy with a thousand tasks, flitting everywhere, above, below, on every surface vertical or horizontal, and in two long lines on either side of the gray mass. Yet, from this distance, there is no sound.

What the scaffolding conceals is a ship, and this ship already has a name—any one of several, according to whom you ask. To the more than five thousand electricians, pipefitters, welders, braziers, tinknockers, riggers, anglesmiths, straighteners, blasters, and shipfitters who labored out on the deckplate to put it together piece by piece, and to the legions of naval architects, draftsmen, and marine engineers who designed its parts and supervised its construction, it is Hull 463, the four hundred and sixty-third vessel to slide down the ways of Bath Iron Works since the yard's founding in 1884 by an ambitious local named Thomas Worcester Hyde, a Civil War general and Medal of Honor winner.

To Mrs. Laurette Giroux Cook of Surprise, Arizona, who will crack a bottle of domestic champagne against the gray-drab hull in a matter of hours, the ship's name has a more poignant, and perhaps painful, meaning; she knows it as the USS *Donald Cook,* memorial to a husband and father lost to war, by all accounts an extraordinarily selfless man who died of malaria while a POW in Vietnam.

To the Navy, which will be represented at the launch by Rear Admiral George A. Huchting, this looming hulk of steel is the DDG-75, the twenty-fifth Arleigh Burke–class Aegis guided missile destroyer. The ship's crew—the three hundred and fifteen enlisted sailors and twenty-two officers, some of whom first assembled in solemnity for the laying of the initial wedge of her curved keel, who let fall their scribbled hopes and prayers beneath the foot of the ship's aluminum mast as they helped to raise it up—will of course quickly come to know her as intimately as the men who built her. For the next thirty-five years, as tens of thousands of miles of sea pass beneath her hull, she will be to some a prison, to others a home and refuge, but surely to all a place of the endless boredom of watching and waiting punctuated by a few hours or days, perhaps, of the frenetic, adrenalin-driven action for which she was built—war.

Hull 463, Arleigh Burke destroyer, the USS *Donald Cook*, DDG-75—the particular name is not really all that important. It is a ship, assembled over four years piece by piece, steel plate by steel plate, from the first half-moon slices of keel to topmost radar mast, almost by hand. As Allan Cameron, president of Bath Iron Works, will tell the gathered dignitaries later in the morning rain, this ship will be launched following principles not much evolved from the time of the world's first great seagoers, the Phoenicians, thirty-five hundred years ago. And all that is about to change.

Across the quiet water and just upstream from the industrial cacophony of Bath lies the town of Woolwich, its poorer, dilapidated houses lining Route 1 giving way to riverside estates and rich spreads tucked away in the folds of a rocky landscape that rises and falls in gentle hills carpeted with the green-gray blur of mature pines. On the Bath side, above the bridge, a handful of unremark-able boats are tied up at moorings in the river, modest powerboats mostly, twenty to forty feet long, good for cruising down the river to Casco Bay, fishing for blues or mackerel or stripers and drinking beer from cans while the sun beats down. The trees and the boats at rest, these are two of the three foundations of Maine's economy: the lumber, pulp, and paper companies that own vast stretches of the northern and inland regions of the state; and tourism, from the snarl of winter's snowmobiles to the crack of lobster claws yielding their flesh to summer's hungry visitors, up "from away."

Maine relies on the vagaries of resource extraction on the one hand, while ingeniously exploiting the enjoyment of those same resources by fickle crowds of visitors whose travel decisions turn on a nickel change in the price of a gallon of gas. The third unsteady leg of this tripod is made up of a handful of industrial giants like S.D. Warren Paper and National Semiconductor. Bath Iron Works is the largest, employing nearly twice as many people—7,800—as the runner-up, the catalog and retail outfit L.L. Bean. With a $300 million annual payroll and tens of millions more paid out to suppliers and to towns and the state in property and income taxes, BIW is the engine that drives midcoast Maine's economy.

A plan has been proposed to create a huge visitor and convention center built around a marina where those few boats now drift quietly with the tide, a plan whose popularity inevitably ebbs and flows with the shipyard's uncertain fortunes. Naval appropriations and military base closings are always front-page news in the *Times-Record,* the local daily paper published in Brunswick, the next town up the road, and always the topic of conversation over coffee in the morning and beers at night at any one of the numerous inexpensive bars and small restaurants that face the yard under the elevated lanes of Route 1, which splits the town in half.

Two Mainers meet for the first time. "I work in Ba-ath," says the first, the long, flat "a" of the coast accent rendering the name distinctly British. "Whatcha do at the shipyahd?" the other asks. Town and yard are one, and as every city councillor knows, as goes BIW, so goes the town.

BIW's economic pull is so strong that the merest ripple of bad news at the yard sends waves out into the surrounding communities. When union contracts come up for negotiation, those waves are felt from Portland to Augusta, where car dealers and building supply stores, boatyards and real estate agents all feel the pinch of unspent dollars hoarded against a strike, like the one that crippled the region in '85. Everyone who lives here knows that building ships for the Navy is an uncertain game, yet it is still the only game in town. BIW, like every other shipyard in the United States, simply cannot compete with the massively subsidized shipbuilders of Korea, Japan, Russia, and other countries when it comes to the commercial shipbuilding business.

The people of Bath know that their town occupies a very special—and vulnerable—position as home to one of the six active naval shipbuilders left in America. What they may not know, or want to think about, is that there were thirty just fifteen years ago, and two more are projected to close for lack of contracts by the year 2005.

To survive through the turn of the millenium, BIW must modernize, and that means turning its back on the old traditions. No more ships built on inclined ways. No more launches with the great gray leviathan rushing down over concrete made slick with wax and grease to splash into the Kennebec. Three, perhaps four more hulls, and all that will be finished. It may be finished forever.

Standing high above the river in the middle of the night, it's all too easy to imagine the ships departing, the lights winking out one by one, the buildings rusting away in the salt air. It happened once before, in a flurry of stock market speculation and naked greed that began in 1905 with Charles Schwab, J. P. Morgan's right-hand man, helping to orchestrate what today would be called a leveraged buyout. It took nearly twenty years, but the accumulated debt finally caught up with the company, which went bankrupt in 1925. The yard lay empty, dead, until a former BIW marine engineer, William "Pete" Newell, bought it and began a slow resurrection based initially on building quality palatial, made-to-order oceangoing steam yachts for the likes of J. P. Morgan and Charles Sorensen, vice-president of the Ford Motor Company.

Were history to repeat itself, it is unlikely that BIW would find a rescuer with the billion or so dollars it would take to make the yard competitive in today's market. Instead of looking down the river at the forest of cranes and staging and hearing the muted bustle of industry, the only view would be to the north and of a different kind of forest, this one of green and white navigation lights from the hundreds of pleasure craft anchored in front of what no doubt would be just one more marina/condo shoreline development, one more destination for the party boats and the out-of-state conventioneers.

Up on the Carlton Bridge in the middle of the night it is easy for the observer to succumb, awestruck at the sheer industry and ingenuity of those who create such things; to build something that enormous, that complicated, seems a miraculous endeavor. Yet it is not this single ship poised for launch on the morning tide that is the miracle. That BIW itself still exists, continuing a shipbuilding tradition stretching back almost four hundred years in this awkward and unlikely place, and at a time when so many forces are arrayed against it—that is the true miracle.

THE YARD

The USS *Donald Cook* tied up at Bath Iron Works' North Pier, summer 1998. The radar mast framed by the cranes to the right is of a ship abuilding on the ways.

THE DDG-51 IN BRIEF

The world changes rapidly. Five years ago, who was talking about our biggest problem being a ten-, fifteen-year-old second-hand diesel sub the Russians sold off to somebody? Now you're trying to chase every third world country that's got a couple of diesels defending their coast.

—Lieutenant Commander Mike Anderson,
Combat Systems Officer, USS Donald Cook

On any typical summer day at the main riverside yard of Bath Iron Works, launching ways, buildings, and piers are all crowded with DDG-51 Arleigh Burke–class Aegis guided missile destroyers in various states of completion. They take their name from Admiral Arleigh "Thirty-one Knot" Burke, a naval hero of several key World War II Pacific sea battles and later three-term chief of Naval Operations, the CEO of the Navy. He was the first man to recognize that the natural speed and stealth of the destroyers could give them a role much broader than that of outriding escorts for carrier battle groups. His nickname came from his insistence on speed, that every ship in his battle group be able to steam at 31 knots for sustained periods of time. The sustainable top speed of the Arleigh Burkes today is about 31 knots.

The Navy classifies its ships in ways not always comprehensible to the layman—or even consistent with any logical system. In "DDG-51," the "DD" designates the class as a destroyer, the "G" as one whose main armament is guided missiles. Yet there exists a previous class of ships, the CGs, or guided missile cruisers, which share propulsion, weapons, and radar systems, and are also very similar in size and displacement. In a further muddying of the waters, the whole class, which may have thirty ships, can be referred to generically either by the number of the first hull (the "DDG-51 class") or by the name of the first hull (the "Arleigh Burkes").

In practice, what you call the ship depends on who you are. Deckplate mechanics in the yard and shops call her Hull 463; the *Donald Cook's* officer and crew call her the *Donald Cook,* or more formally, DDG-75. Knowing that the DDGs started with 51, anyone can determine that the *Donald Cook* is the twenty-fourth of this class to be built.

The "Aegis" in the full name refers to the unique integration of radar and weapons into a single system capable of identifying, evaluating, prioritizing, tracking, and firing on multiple threats in a much more rapid and effective way than previous, more fragmented systems. "Aegis"—"shield" in Latin—also underlines the traditional role of destroyers as the guardians of the far larger and more valuable real estate, battleships and aircraft carriers particularly, especially against torpedo attack from submarines.

In classical mythology the aegis was the shield of Zeus or Athena, made more potent by the head of a Gorgon mounted in its center. Enemies confronting the gods in battle would be turned to stone should they look directly at the shield—and thus into the eyes of the Gorgon. Whether the arms contractor or naval functionary who first christened the system the "Aegis" borrowed on the more poetic interpretation—Athena is a hunter, after all—to suggest the wider role the system might play in attacking versus passively defending, no one knows.

On this muggy August afternoon the USS *Donald Cook* is tied up alongside Pier 1, the barrel of its five-inch deck gun aimed as if to take out the main span of the Carlton Bridge just a few hundred feet upriver. On the angled launching ways, the *Oscar Austin* looks eager to plunge into the river flowing swiftly by at its stern, while the *Winston S. Churchill* beside it has a complete keel and awaits all of the units of upper decking and superstructure. A half mile south down the river, Pier 2 is home to the nearly complete *Higgins*, waiting to travel to BIW's Portland drydock to have its protruding sonar dome installed.

At the south side of Pier 3, still called the New Pier though it hasn't been new for decades, the *O'Kane* is made fast, a ship in that awkward adolescent stage where the blemishes are so overwhelmingly visible that it's hard to imagine maturity just around the corner. A passerby could be forgiven for thinking, at first glance, that she was home for a refit after a battering at sea rather than months away from sailing off for her first sea trials. Her hull and bulkheads are a messy collage of green, orange, and haze-gray paint, with gaping holes in bulkheads and deck yet to be sealed over. Dozens of interior spaces for machinery, electronics, and berthing are empty and unfinished, while bare foundations small and large protrude from every horizontal and vertical surface on deck awaiting lifeboats, weapons, and communications and navigation equipment of every variety.

In the Hardings Fabrication Facility three miles up Bath Road towards Brunswick, and in the East Brunswick Manufacturing Facility just across the street from Hardings, two shifts run from 6 A.M. to midnight, Monday to Friday, with overtime either before or after the regular shifts and Saturdays and Sundays besides. In the hangar-like Assembly Building behind the ways, the work of shipbuilding goes on twenty-four hours a day, three shifts, seven days a week when necessary. Steel plate is being burned and welded up, pipe bent, sheet metal formed into ducting, small parts fitted into larger ones, and hull plates shaped for BIW Hulls 470 and 471, destroyers that do not even yet have names.

On the other side of Brunswick, meanwhile, ten miles up the road, engineers, naval architects, and designers labor assiduously on Naval Sea Command's never-ending modifications and updates to a hull now in its fifteenth year of production.

The *Donald Cook* is the fourteenth destroyer of its class to be built at the yard; Litton/Ingalls Shipbuilding in Pascagoula, Mississippi, has built a further eleven. Taking into account future plans for more hulls well into the first decade of the twenty-first

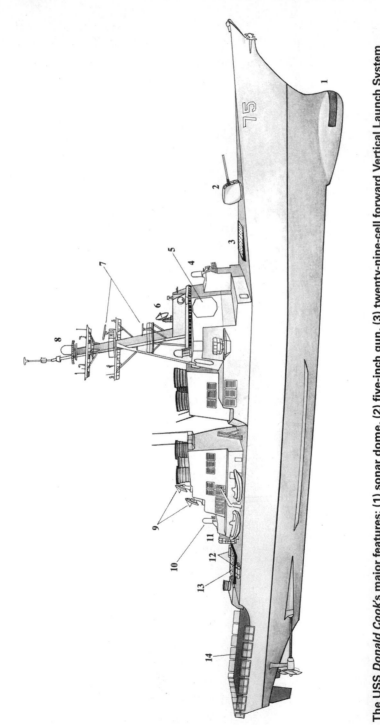

The USS *Donald Cook*'s major features: (1) sonar dome, (2) five-inch gun, (3) twenty-nine-cell forward Vertical Launch System (VLS) missile launcher, (4) forward Phalanx CIWS, (5) SPY radar arrays at each corner of deckhouse, (6) forward fire control illuminator, (7) navigational radars, (8) Identification Friend or Foe radar, (9) aft fire control illuminators, (10) aft Phalanx CIWS, (11) Harpoon missile launchers, (12) port and starboard torpedo launchers, (13) aft sixty-one-cell VLS missile launcher, (14) fantail.

century and the planned decommissioning of the cruisers and frigates that overlap the DDG-51's strategic role, the result is that the Arleigh Burkes not only are the largest naval shipbuilding program of our time but are soon to be the largest—and most visible—class of ships afloat.

The DDG-51s make up, by all accounts, a very special and unusually well loved class of hulls. "A sweet ride . . . just a gorgeous hull . . . a superb piece of engineering and shipbuilding . . . a brilliant, brilliant hull design," say the officers who command them and the crews who man them.

The ship is 504 feet long overall, almost 67 feet wide in the beam, and draws about 30 feet of water. The main deck, where you would be if you boarded at the fantail, is called Level 1, with three more decks below: Levels 2, 3, and 4 (which is the lowest). Up from Level 1, there are five decks (Level 01 being the first above Level 1) with the bridge at Level 05 about 55 feet above the waterline. The very top of the single back-raked radar mast is nearly 168 feet off the water. Four 25,000-horsepower gas turbines drive two propellers for a cruising speed of 16 knots and a top speed twice that, or approaching 40 miles per hour.

In appearance she is very sleek and spare, with her superstructure, mast, and twin stacks angled back, her bow rising steeply from the water, her fantail flat and wide enough to take an incoming Seahawk helicopter comfortably. The decks are uncluttered, and unlike the popular war film image of the battleship bristling with guns, the DDG-51's visible weaponry is deceptively minimal and unobtrusive.

Forward, a single five-inch deck gun squats behind the anchor capstan, and behind it and nearly flush to the upward-sloping deck is the waffle grid of missile silo hatches covering the twenty-nine cells of the Vertical Launch System (VLS). Two levels up, just below the windows of the bridge, what the ensigns call the "see-whiz"—the Phalanx CIWS (Close-In Weapons System), a rapid-fire Gatling-type cannon—stands upright, an innocent-looking white dome the height of a man. It is a last-ditch defensive weapon which, when an incoming threat, most likely a missile, is detected already so close that the ship has no time to get its own missile off, will fill the air with a curtain of depleted uranium

or tungsten projectiles meant to shred the nose cone and deto-
nate the incoming threat before it hits the ship.

Aft of the stacks just inside the rail on port and starboard are
Harpoon missile canisters. Two clusters of deck-mounted torpedo
tubes, three per side, flank the sixty-one-cell aft VLS, while the
rear approaches to the ship are protected by another Phalanx set
three levels above the fantail. The ninety-cell VLS together can
fire four kinds of missiles: standard missiles (SM-2s), medium or
extended range, designed to bring down hostile aircraft; anti-submarine rockets (ASROC); and two types of Tomahawk cruise missiles—TASMs, or anti-ship missiles; and TLAMs, or land-attack
missiles, such as those directed at Iraqi targets during the Gulf
War and used most recently against targets in Yugoslavia. Some
weapons officers have been known to say that, contrary to her
benign silhouette, the DDG-51 is in fact the most heavily armed
ship, ton for ton, in the whole Navy.

Though a cruiser captain might debate that, what does set the
Arleigh Burke apart from other "platforms," as the Navy calls
classes of ships, is that in 1979 when it was originally conceived,
this destroyer was the first ship the Navy had designed from
scratch in twenty years. Though they were certainly building on
prior experience in World War II and the Korean War, more
recent engagements—our own in Vietnam and the British navy's
costly Falkland Island's sea battles especially—gave naval strate-
gists and architects a foretaste of the world to come.

Lieutenant Commander Mike Anderson is tall and spare, his
hands and fingers in constant motion, his whole presence intense
and focused. By Navy career path he is an EDO, or engineering
duty officer, one of the land-based managers who run the shipbuilding programs, determine future production and procurement, and make critical decisions about combat systems, logistics,
and platform capabilities and limitations. His assignment to the
Donald Cook as CSO, or combat systems officer, is something of
an experiment.

With eleven years in, and more than two-thirds of that time
with the surface fleet, Anderson is part of the Navy's attempt to
bring active surface duty officers with hands-on experience of the

ships at sea back to land to apply that experience to the next generation of warships.

We are fleet guys, Anderson says, who are able to focus on learning the shoreside infrastructure. The surface guys need to get back out to sea. We aren't worried about getting through the inspection in two weeks; we are worried about how to fix the problems out there in the next generation, in two generations . . . the long-term focus. Admiral Minot, who used to be the head of the engineering-duty community, had a quote that I liked. *We are the conscience of the Navy,* he said.

When the Navy went to the Arleigh Burke, Anderson continues, they went from the ground up. It's all steel. You don't have the aluminum superstructure like you did on the earlier frigates and cruisers. A lot of lessons were learned, a lot of problems from the previous classes were finally wrapped in and fixed in the design of this ship.

The particular problems he refers to have names known to every naval officer from Virginia to Vladivostok—the HMS *Sheffield* and the USS *Stark,* the latter of which was built with an aluminum superstructure in an era when armor protection was sacrificed on the altar of weight reduction for greater stability and cruising range. During the Falklands War, the Argentines hit the British destroyer with two Exocet missiles; the Iraqis disabled the *Stark* with two of the same French-designed missiles during the Iran-Iraq War.

In neither case did all of the missile warheads explode. Instead, the violent impact of the direct hits blew quantities of unexpended rocket fuel throughout the ships, causing ferociously hot, uncontrollable fires. Twenty men died on the *Sheffield,* which could not be salvaged, a significant loss of prestige as well as of human life for the British navy. The attack on the *Stark* cost thirty-seven lives and ended the careers of several of its officers, who were judged to have failed to keep the ship and crew in a sufficient state of preparedness. More important, both of these incidents demonstrated once and for all that a single low-cost missile commercially available to any third world country could put a billion-dollar warship out of commission, if not destroy it completely. One part of the problem was certainly in the material used during construction.

With aluminum, Lieutenant Commander Anderson says, you have corrosion problems obviously, damage control problems, especially if you get fires burning aboard. Aluminum has a lower combustion temperature. Your biggest fear on the water is fire, your worst fear, you're out there and you're in a no-win situation.

The Arleigh Burkes are all steel but for parts of the funnels and the radar mast. The belowdecks critical spaces are protected by, in some cases, double bulkheads of steel, which tend to absorb and deflect the blast of an incoming missile, while some above-deck spaces are armored with 70 tons of Kevlar.

When asked about the mission of the *Donald Cook* and the other DDGs, Anderson laughs and throws up his hands. Last week, today, or tomorrow? he replies, only half-joking. That's really the value of these ships—they can perform virtually any mission, and do on a daily basis. The *Laboon*, DDG-58, was firing Tomahawk cruise missiles in the last round of the Gulf War, launching strikes on Iraq. So you've got a strike capability. Every battle group takes along two Aegis destroyers, usually with two Aegis cruisers, so they are an integral part of the battle group air defense.

With the advent of the next generation of standard missiles that will be able to fire off the destroyer as well as the cruiser, he continues, you are starting to talk about endoatmospheric theater ballistic missile defense—area defense. So we are able to defend a beachhead from inbound missiles, and Iraqi Scuds, North Korean Nodongs; some of the shorter-range, two-hundred- to four-hundred-kilometer stuff is right in our envelope.

The reality is, Anderson says emphatically, we are not going to be out there in a hot shooting war. We are not going to be in a long, protracted, World War II–type battle where, if you see the bad guy, you shoot him. That's just not the way the world is struc-tured right now and that's not the way the conflicts are going. It's going to be short, high-intensity; it's going to happen quick, with-out warning.

CIC, the Combat Information Center, is the place from which war is waged on the *Donald Cook*. It is a large, square, low-ceilinged room illuminated only, it seems, by the soft green and orange light of glowing computer consoles in front of which sit

The tactical actions officer's station at the front of the Combat Information Center. To the far right are the Tomahawk controllers, and at the back of the room, air alley. Ceiling and walls (as well as the masses of cabling overhead) are black-painted, and the room normally is dark, to better enable the controllers to read their screens.

the score or so of highly trained sonar, radar, weapons, and fire control operators in comfortably padded chairs. The front of the room is dominated by a flat, orange large screen display. The LSD shows the big picture—the location of every ship, for example, in the entire area the ship is operating in—for the benefit of the captain and the TAO, the tactical actions officer, whose chairs face it directly. With the ship anchored in Bath, the whole East Coast of the United States is lit up in outline, with the location of the Donald Cook picked out as a blip.

CIC is organized by warfare area—electronic warfare (detecting, identifying, prioritizing, tracking), systems (monitoring the functioning of all combat, weapons, and communications systems), air warfare (engaging airborne targets), subsurface warfare (engaging underwater targets), surface warfare (engaging floating targets), and strike (engaging targets inland from offshore). The three sides of the room facing away from the front screen are lined with consoles, each dedicated to some aspect of one of those five tasks.

From one corner the radar operator can direct the four SPY arrays, which are the eyes of Aegis, to pay attention to a particular area or to scan a target more or less frequently (from four to sixty times a second) depending on its assessed risk. A systems operator constantly checks and updates the status of weapons, communications, and all primary electronic systems. The IDS, or identification supervisor, helps determine if acquired contacts are friend or foe.

Along the back wall, what they call "air alley," the first chair is occupied by the missile systems supervisor, who's controlling the launchers and the CIWS, among other things; and next to him sits an anti-air warfare coordinator. Because the primary mission of destroyers has always been to protect the aircraft carriers by screening the perimeters and keeping air and water safe around them, the DDG-51s, though they don't launch planes, may direct patrol aircraft off other ships. Along air alley, there are two such AICs, air-intercept controllers, one tasked with shooting missiles if necessary at hostile planes some ways away from the ship, another controlling flights of F-14s and F-18s in carrying out the same defensive task.

Next to the AICs is a submarine air controller directing helicopters and anti-submarine aircraft, then a surface warfare coordinator, and finally consoles directing the targeting and firing of torpedos, a station for figuring vertical launchpad and ASROC solutions, which program missiles for the best way to get from ship to target. The right wall is taken up by the five-inch gun controller, the Harpoon console, and the Tomahawk missile console.

In the first chair along air alley sits FC-1 Tim Gilmore, a fire controlman first class, who is responsible for firing all of the different missiles out of the VLS launchers and controlling the CIWS. A tall, slim redhead with an easy grin and a gentle North Carolina accent, Gilmore, with eleven years on the system, considers himself one of the old men among Aegis operators. In order to be an Aegis fire controlman when I came through, he says, you had to be the best of the best.

Aegis, he continues, is not your normal turning radar. We

don't have to wait for that antenna to come around on its sweep in order to see any point in space at any given time. All we have to do with Aegis is point a beam there. We are seeing 360 degrees around the ship every second or faster from zero to sixty-four nautical miles out. Aegis will automatically, if it sees a contact out there, get the contact's range, course, bearing, and all that goes with it, and report it to another computer, the Command and Decision computer, which displays this on our scopes as symbols.

Depending on the altitude of planes, Gilmore continues, we can see them two hundred and fifty nautical miles out. We'll tell the system, if you see a [contact] that's flying below five hundred feet, doing better than Mach 1, and it's coming inbound to the ship, that's a threat. The system will automatically assign it top priority. It's going to look at it several times a second to update its track so we can get as accurate a fire control solution on the target as we can.

The WCS [Weapons Control System], Gilmore explains, handles all the number-crunching involved in the fire control solution—in predicting, when I fire a missile now, at what point the target is going to be in space given that it takes the missile X amount of time to get there. The WCS does all the weapons assignments; it will say, I want this particular kind of missile in this particular VLS cell to go to that bird [plane] or to that target.

Initially, the radar controller, he continues, is going to pick the target up on radar. The ID controller will do IFF [Identification Friend or Foe] interrogations on it and try to identify it as a friendly track. An IFF challenge is basically a coded radar signal that we send out. All commercial airliners, all NATO aircraft, have IFF aboard, and they've got a transponder that will receive the interrogation and reply. We'll use assets—say, fighter aircraft—to go over and visually identify the target. If we decide this is a bad guy and we want to hit him, the anti-air warfare coordinator is going to select that track, tell the system to engage it with a standard missile, and then the engagement comes here, to my console, missile systems supervisor.

I will have FIRE AUTHORIZE blinking at me, and I will fire authorize—the missile will leave. The SPY radar is going to be talking to the missile the whole time, giving it course correction updates to where the target is until the very last few seconds in flight. Then our target control directors [also called fire illumina-tors]—those big dishes on top of the ship—will shine radiation at it. The missile is going to follow, not the radiation off the dish, but the reflective radiation off the target.

The biggest advantage is, Gilmore says with a broad smile, as soon as we turn on this illuminator—you see it in the movies—their cockpit goes crazy. Missile lock-on, aaaahh! Gilmore yells, throwing up his hands in mock panic. That's when their threat receiver starts up [telling the pilot with both a visual and an audio display] because it sees our fire control radar. In the old days, we used to have the SM-1 missile, and it rode the beam all the way. The entire time our missile was in flight, the target knew he was being engaged. Now, with the later missile upgrades and the Aegis ships, the target is only seeing a couple seconds of, *You're engaged!* before the missile is *there*.

The biggest problem today, Gilmore points out, is identifying someone in time before they do something to us. If a plane is inbound, generally we will give them a Level One warning around fifty miles from the ship. We are going to go out over international air distress and military air distress, both those circuits and say, *You are approaching a U.S. naval warship operating in interna-tional waters. State your intentions.* If they continue inbound, somewhere around thirty nautical miles from the ship, we will give a Level Two warning. *If you continue on this present course, you may be fired on.* If they still continue, we light them up—hit them with the fire control director that will set off their threat receivers. We give them those three chances. In today's day and age, we will not fire on anybody unless they fire on us.

In a time of war, Gilmore notes, that all changes. We are to take any action necessary to defend the ship. We are not to fire on anybody unless they fire on us first. In the case of an Exocet mis-sile, you see the target split off from the Archer—we call the fir-ing aircraft the "Archer." The Exocet has a little cruise phase, and then it's going to drop to the deck. At Mach 1, a little under 750 mph, it's a relatively slow missile. The problem with it is, it's a very low flyer. It will get down under the deck and hide under the cur-vature of the earth, so you won't see that missile until it pops up over the horizon probably fifteen nautical miles [from the ship]. We probably have forty-five seconds, which is, as far as Aegis is concerned, an eternity.

We'll see the Exocet drop down, and we could get a missile off on it very easily, but then we are going to lose track on it [once it disappears below the curvature of the earth out of the reach of the ship's radar]. So we aren't going to fire. We are going to wait. The way most ships operate in the Navy, the first thing we are going to hit, the first priority, is an inbound missile. Now, if somebody is firing on us, it really depends on what they are doing. If they fire and turn out—away—and they are no longer a threat, we may not do anything. We'll take out their inbound missile, but they are no longer a threat.

It depends on the current geopolitical situation, Gilmore says with the shrug of someone who has sat in a ship off the coast of Kuwait anticipating just such a situation. If we have been having small hostilities with the Iraqis, and they just kicked our weapons inspectors out of the country, and said death to all American pigs, we may immediately take the Archer out.

In any case, the missile, we'll lose the track on it, but the Command and Decision computer will keep updating from its last known course and speed, what we call a "coaster." So we know what area we might expect to see the missile in, we're projecting its plot. The SPY radar is automatically going to be looking right in that area. The SPY operator will also do a SPY acquire, [direct the system] to look at a certain spot three or four times a second, looking for any contact in that area. In real life, we know we have a missile inbound, we saw the target split, we know it dropped to the deck. We are going to see it in plenty of time.

The way we do our thing in combat, the tactical actions officer will tell the anti-air warfare coordinator to assign track number 80564, and he'll say, *Air Attack, kill track 80564.* The air operator will say, *Air, aye, killing track 80564!* and select that specific track [on his screen] and take it into close control. He will tell the system, ENGAGE STANDARD MISSILES. Then the Command and Decision computer tells the Weapons Control System, does the number crunching, assigns a missile, assigns an illuminator, then alerts the missile systems supervisor, me. This console will squawk at you—

you get a beeping in your ear that's an alert. You press REVIEW ALERTS and up pops this alert which tells you what this is all about. I would get something that says RECOMMEND FIRE. I would press a button that says FIRE AUTHORIZE. As long as I have heard the TAO tell the air operator to kill that track, I don't need any further action.

Gilmore was on an Aegis cruiser during Desert Storm, his first time in combat. He can't tell you if he did push that button or what the circumstances were. I am a Desert Storm veteran, he says simply. There was a lot of fire in the Gulf, and I take pride in wearing this uniform.

The unusual conditions of the Gulf War showed just how adaptable the DDGs were. We were originally designed and built for the Great Red Menace, Lieutenant Commander Anderson observes. The Aegis system was originally to stop giant surface-to-surface missiles and air-to-surface missiles coming in screaming at Mach 2, Mach 3, long-range cruise shots from Soviet aircraft—the Backfire Bombers and Sovremenny guided missile destroyers. Our anti-submarine systems were geared the same way—to detect, localize, and identify Soviet nuclear attack boats.

They're not the bad guys anymore, he points out. We are not worried about the super-sexy Russian nukes—the Victor 3s and the Akula (Shark) subs. Now what you are trying to stop is the French Exocet missile, the Italian Automat missile, and other lower, slower, smaller missiles coming in from closer range, from patrol boats close in off the beach. Now you are operating close to the beach with a lot of radar clutter.

The world changes rapidly, he says, shrugging. Five years ago, who was talking about our biggest problem being a ten-, fifteen-year-old secondhand diesel sub the Russians sold off to somebody? Now you are trying to chase every third world country that's got a couple of diesels defending their coast. The real key to the DDG is its flexibility, its ability to adapt, he finishes.

Since the commissioning of the first Arleigh Burke–class DDG-51 in 1991, these destroyers have indeed proven to be remarkably adaptable. They have patrolled in defensive and escort roles in the Persian Gulf, launched cruise missiles in the Desert Storm conflict, and been a significant part of the Navy's deterrent presence off the former Yugoslavia and Haiti during those conflicts. Two Aegis destroyers accompany every battle group in waters around the world. Most recently, they have played an active role in the Kosovo conflict.

I think, Lieutenant Commander Anderson concludes with some force to his words, the Navy in general and the Arleigh Burkes in particular have kept us out of a lot more trouble than they've gotten us into. They [help] create a great wealth for the nation. They create a great safety and stability for our nation. And they've enabled us to create stability for others. You look at Asia now and the enormous economic boom that's taken place over the last twenty to thirty years there. That was brought about by stability. Without foreign capital you don't have economic growth. Without stability you don't have foreign capital. It's a worldwide economy. And the ability to support peace around the world, to increase stability, saves lives, promotes quality of life, human rights, and the rest.

A STEEL SHIPBUILDING PRIMER

Ship design is fascinating—the ship is a city, and, oh, by the way, it's got to go 30 knots, too. Naval architects have to do everything land architects have to do, and yet the ship's got to move and survive in hostile environments as well.

—Russ Hoffman, Manager of Concept Design Development, Bath Iron Works

The Church Road Office Facility—CROF, for short—is where Bath Iron Works does a significant part of the design and engineering work on the ships it is building, has contracted to build, and even would like to build if only someone would agree to buy them. It is in a semi-industrial part of Brunswick just off I-95, after the Vo-Tech school and before the Mariner's lumberyard and sawmill, one of those extraordinarily ugly, low, brick and concrete-block buildings from the '60s when flat roofs were in and everything looked like a high school built after a compromise with overtaxed parents.

Inside, the decor is one more testament to the fact that BIW does not waste precious capital coddling its employees with plush surroundings. The lighting is 100 percent fluorescent, the walls and floors of colors not so much soothing as drab to the degree of invisibility. The metal desks could be government-issue while the floor plan is your basic acres of cubicles in one vast room surrounded by cubbyhole offices and conference rooms. Overall, the effect is remarkably devoid of perspective, so that the visitor has the impression of having stepped into just one more of the conceptual drawings that these hundreds of busy minds work so hard at producing every day.

At every desk is at least one computer, behind every computer an engineer, around every engineer's neck a blue or black ribbon with BIW ENGINEERING picked out in white letters, and from which dangles an ID badge. Unlike almost every other place in the yard, these guys have short hair, clean shirts, and few visible tattoos. The atmosphere is hushed and studious, with the intensity of a library or a museum, a place where a lot of brainwork gets done and people need quiet to think.

Four bright blue-and-white signs hang from the ceiling, dividing the central workspace into strategic teams—signposts to BIW's past, present, and future. One team labors on the ever evolving DDG-51s, whose latest version, Flight IIA, has been significantly modified to house, fuel, and arm two Seahawk helicopters out of a hangar on its stern.

A second sign represents what might have been: the arsenal ship, a radical, low-cost Stealth platform that would have delivered up to five hundred missiles into a war zone, yet be manned by as few as fifty crew. BIW was one of three consortiums chosen for a second round of design competition, but the project was abruptly defunded, a victim of clashing interests within the Navy and the Pentagon.

A third team works under the LPD-17 sign, which should be larger than the others, for it represents a tangible triumph: BIW's selection over rival Ingalls Shipbuilding to design and build an Amphibious Assault ship together with Avondale Industries in Louisiana. Avondale will build the first two, with BIW slated to follow on with every third ship towards an as-yet undetermined total. Actual construction is a scant two years away, and this group is busy.

At 684 feet in length, 105 feet at its widest, and displacing 25,300 tons at full load, the LPD-17 is far larger, wider, and heavier than anything BIW has built for two decades. Its interior spaces are meant to carry things—Marines, vehicles, ammo, matériel, fuel, helos. Such spaces require an extraordinary amount of steel fabrication which, given the yard's limitations of available manufacturing capacity and space, is going to present a large challenge.

On the final sign is written "SC-21," four deceptively simple characters that represent the long-term future of the yard. Surface Combatant 21st Century is, to most every military shipbuilder in the country today, the pot of gold at the end of the rainbow, or at the very least, one huge pie from which they hope to be served even a small piece. Though at the most preliminary of stages—the Navy has not yet even put out a request for proposals (RFP)—it promises to be a technologically innovative, multimission, high-complexity destroyer-type platform. In short, it is exactly the kind of ship BIW builds best.

Since 1985, when BIW won the contract to build the lead—the first—DDG-51, the Navy has come a long way in its thinking about how to get the most ship for the least money while still keeping enough shipyards alive to fill its future needs. Today's approach, to simplify grossly, is to describe in brief outline what they want and let teams, usually composed of a shipyard together with electronics, communications, and weapons manufacturers, compete to design the best and least expensive ship. Implicit in this is the recognition that each yard will design something it can manufacture efficiently—and therefore more cheaply.

Russ Hoffman is a naval architect who began working at BIW fourteen years ago. His first job was doing the launching calculations for the Aegis cruisers BIW was building in the mid-'80s. Today he is the manager of Concept Design Development, a new and still small working group that radically departs from BIW's usual way of doing business. The goal of Hoffman's group is, instead of waiting for the shipbuilding customer to come to them or request a proposal, to take some very basic design requirements and generate a ship design concept for the potential shipowner.

Hoffman is a soft-spoken and unpretentious man, but brimming over with energy and enthusiasm. With a slightly rounded face, soft brown eyes, and light brown hair going elegantly gray at the temples, he resembles Jeff Bridges, especially in the slight huskiness of his voice, which makes him sound always on the edge of a good chuckle. It is not hard to imagine him shmoozing equally well with

Navy brass and hardheaded CEOs—the only two groups of people in the world, after all, who have the hundreds of millions of dollars it takes to build a single ship. He has put in enough time in various engineering capacities to know the acquisition and design process from top to bottom.

In the traditional Navy acquisitions program, he explains, which we could say the DDGs were, we received from the government a thick specification, probably two three-inch-thick binders full of specs, and a couple of file cabinet drawers full of contract drawings and contract guidance drawings. With those, we were to proceed to do enough initial engineering work to put a price on the ship and make a bid.

So, at a relatively high level, he continues, the ship was designed when the government asked us for a price. There's still a lot of detailed design work to do, but we know what the scantlings—the raw dimensions—are, we know the requirements for whole systems. They told us what engines to use and what propellers and reduction gears. So we are not strictly responsible for performance. They told us the hull shape, the engines, we put it together.

It's the distributive systems, he continues, the cables, the vent ducts, the piping, that we had to locate and properly size and support. That's our job. It's also our job to go through the structure to make sure the scantlings they've given us are right. Now, government Navy contracts have a weight and a center of gravity penalty clause that says if you exceed the weight you sign up to in the contract, then the penalty is $250 thousand per 10-ton increment over that limit; and the center of gravity height, which is of course critical from a stability point of view, is $1.25 *million* per tenth of a foot over that limit.

To the uninitiated, a penalty for building an excessively heavy ship makes sense; more weight means more fuel means less room for weapons/men/matériel and less speed, and thus perhaps the ship's intended mission is compromised. Understanding center of gravity is more difficult, though it begins with the simplest of questions: Why do ships float? What keeps them upright and stable in the water during, say, a ferocious storm?

Things more dense than water sink: a rock, a block of concrete, a bar of steel. Things less dense than water float: a block of wood, a buoy, the Titanic before flooding. Steel ships float because they are filled with air, whose density, when averaged with the steel, makes the density of the ship as a whole much less than water.

The Navy requires much more of the DDGs than mere floating, however; they must be stable in high seas and at high speeds, in high winds, when fully loaded, and when in light ballast only.

The location of the ship's center of gravity is so important because it has everything to do with the ship's stability in the water under all sorts of conditions. To understand why, first consider the basic forces acting on any ship in the water. Every object has a center of gravity, the point at which the object's weight can be said to be concentrated by the earth's gravitational pull. The center of gravity of a ship never changes, no matter where it is on the surface of the earth. Ships are designed with a low center of gravity so the keel stays pointed down and the top of the ship remains upright.

Buoyancy is another force acting on a hull: a ship's buoyancy is equal to the weight of the water it displaces, and the center of buoyancy is considered to be the point at which the force of buoy-

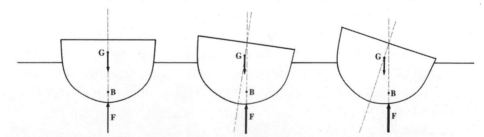

The center of gravity (G), the center of buoyancy (B), and the force of buoyancy (F) at work in a ship on the water. At left, the force of gravity and the force of buoyancy are equal, and the ship sits straight up and down in the water, with the force of buoyancy acting directly under the keel. Middle and right, the ship begins to heel, and the center of buoyancy begins to shift outboard. The force of buoyancy moves in the direction of the tilt as more hull is exposed, tending to right the ship. A blast of wind against the exposed hull, shifting weights on the ship, an inflow of water on one side— all of these can overcome the ship's tendency to right itself.

ancy acts. The force of buoyancy can be thought of, in a sense, as trying to push the hull out of the water, a force anyone who has ever tried to submerge a piece of wood of any size has felt. Another way of thinking about this is that all of the air trapped inside the ship is lighter than the water. Begin filling the ship with water, which displaces the air, and at some point it will sink.

The center of buoyancy on a ship shifts from side to side and up and down as the hull tilts to one side or the other and as waves rock a ship up and down, thus submerging more or less of the hull area. The force of couyancy is therefore not uniform at every point on the hull. Ships do not naturally stand straight up and down, 90 degrees to the water. They list or heel over, for whatever reason, all the time. What keeps them upright and afloat is a low center of gravity—the center of the earth wants to pull the heaviest part of the ship, somewhere low in the keel, toward it at all times, thus keeping the hull down. The streets of many a New World city were originally paved with Old World cobblestones carried in the keels of empty wooden sailing ships to give them sufficient ballast. Lead keels perform the same function on sailing yachts, their weight providing a counterforce against the action of wind on sail, which wants to push the boat over.

On the DDGs, the ballast consists of the hundreds of thousands of gallons of fuel oil, JP-5 jet fuel, freshwater, and the seawater which is drawn into the fuel tanks as fuel is drawn off for consumption. These tanks, called *inner bottoms,* are in the lowest part of the ship, next to the keel, running nearly the whole length of the ship except at the extreme bow and aft, where the prop shafts exit the hull under the fantail. The great weight of ballast is thus concentrated low in the ship and more towards the center.

Every ship has a degree of list beyond which it cannot recover, cannot right itself. Take the *Donald Cook* after launch, floating in the river. The launch weight is relatively light because the ship isn't finished, or fully fueled and watered. A light ship rides high out of the water, presenting more surface area to the wind. The ship gets hit by a blast of wind and begins to heel, to tilt sideways. Two things are happening to keep the ship from tipping over and in fact to give it the tendency to right itself. The force of gravity wants to push the keel back down, concentrating all of that weight closer to the center of the earth. The force of buoyancy, meanwhile, has increased on the opposite side of the boat because more of the boat's hull—and thus the less-dense air inside—meets the water. The force of buoyancy pushes the hull up and against the direction of the wind while the force of gravity pushes the keel down and in the direction of the wind. The net effect is that the boat is always trying to right itself—to return to a straight up-and-down state.

A ship's center of gravity is thus key to its stability, and it is largely determined by the absolute weight of a ship and the distribution of that weight throughout the ship given any particular hull design. Though the government gave BIW its own weight and center of gravity estimates as part of the request for proposals leading up to the bidding for the first DDG-51, BIW must, in a sense, redesign the ship enough to come up with their own estimates as the basis of final negotiations so they don't end up shooting themselves in the foot—paying penalties on estimates they haven't generated.

In the case of the DDGs, the weight and center of gravity estimates involved about thirty people working full-time for several months just to arrive at the rough figures. This document, at several thousand pages, is almost the longest part of the bid, second only to all of the financial scenarios the yard proposes according to whether or not they are the lead contractor building the first ship or the follow-on yard constructing additional ships, how many ships they're to build, and over what period of time, among other variables.

Though Russ Hoffman won't attempt to quantify the number of hours going into the initial design and proposal stages, he does give some indication in a roundabout way. The most recent update of the DDG-51 is Flight IIA, of which the first ship is DDG-79, the USS *Oscar Austin,* which was launched in November 1998, four ships later than the *Donald Cook.*

The IIAs incorporate a number of major changes, the most visible of which is an aft helo hangar together with a tractor system to move helos in and out of it. Of course, there are also facilities to service, fuel, and arm the helos (two Seahawk SH-60R LAMPS III helos equipped for sub hunting), as well as the additional twenty-one officers and crew attached to them. The aft missile deck was raised one whole level, five feet were added to her stern in overall length, and there were other significant upgrades to various electronics packages.

One early engineering estimate, Russ Hoffman says, to do just the design work for the Flight IIA upgrade was several hundred *thousand* man-hours. But that's what it takes to redesign a product of this complexity and make the kind of changes the government wants. When the people at Kvaerner-Masa in Finland [perhaps the most technologically advanced commercial shipbuilder in the world] were advised of that number, they just couldn't believe it. For 900 thousand hours, they could design *five* ships. Their commercial process and their product is so different that it takes them on the order of 200 thousand hours to complete the design!

When it comes to the overall design of the ship, particularly the hull, naval architects talk of primary, secondary, and tertiary stresses. In this case, stresses actually represent design considerations meant to compensate for forces and loading from the global to the local level.

Primary stresses arise from considerations like the *still water bending moment,* a mechanical engineer's snapshot of the gross effect on a structure of all of the forces at work in a ship at rest on the water, arising from its basic hull design. There's a bending of the entire hull that you have to think about first just because of the shape of the hull in the water and the distribution of the weights within it. Because the force of buoyancy is a function of the amount of hull area exposed to the water, ships are not equally buoyant everywhere along the hull. At the bow, for example, which is narrow and steeply arched, there is less buoyancy than at the wide, flat stern whose large surface area exposed to the water means lots of buoyancy.

Large pieces of equipment, or even lots of smaller pieces in one place, can, by their cumulative weight, affect buoyancy (and thus the bending moment) by changing how much hull area is exposed to the water, and where.

A ship hull, Hoffman says, is a big beam, a simple beam with force on it. Because of unequal buoyancy at various points along the hull and the distribution of weights in the hull, there's a bending moment on that structure. It's going to want to sag or *hog*—hogging is when the midship portion goes up relative to bow and stern.

It only makes sense from an engineering point of view, he continues, to see what would happen on a wave; the most critical situation is a wave that is equal in length to the ship. There is a crest at the bow, a crest at the stern, and a trough amidships. You've got more buoyancy at the ends and less amidships, so the ship is going to want to sag. We also look at the case where the crest is amidships, and then the hull is going to hog.

Then there are secondary stresses set up in individual places on the ship, Hoffman adds. Let's take a piece of deckplate that would typically be supported underneath with longitudinal stiffeners.

Now, there's a piece of equipment on top of that deckplate of some weight. There will probably be a web frame [curving vertical

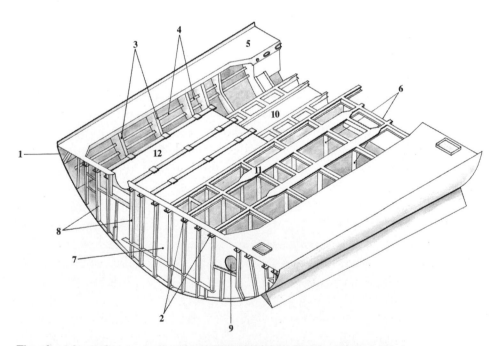

The aft main engine room, Erection Unit 2430: (1) shell plate, (2) deck longitudinal stiffeners, (3) web frames, (4) shell longitudinal stiffeners, (5) deckplate, (6) stanchions, (7) transverse bulkhead, (8) bulkhead stiffeners, (9) Prop shaft cut-out from forward engine room. The two LM–2500 gas turbines will sit on the heavy rail foundations at (12), with the main reduction gear at (10). An Allison 2500-kW generator sits atop the heavy foundations at (11) as well.

ribs of heavier T-beam] at one end and the other, with the stiffeners running underneath. This section of plate and stiffeners between the web frames with the weight on top is going to have its own tendency to sag as well. Same thing with hydrostatic pressure—water pressure—on the panels of plate down below.

So all of these stresses need to be considered when you decide what the size of the steel plate needs to be. We think first about how big the plate should be to accommodate the global stressing of the hull, and then you incorporate into that process the local loading as well.

But depending on where you are on the ship, Hoffman cautions, a local stress might be the governing condition. One of our design criteria is nuclear air blast, what happens when a nuclear weapon detonates at some distance from the hull: portions of the above-water structure have to be designed to take some level of air blast, air overpressure.

There are other reasons for that than nuclear considerations, he emphasizes. We have a gun that sits on the foredeck. Firing that gun creates a tremendous amount of pressure around the barrel that has to be considered. Firing a missile out of the missile launcher, you've got to consider *that* overpressure in the design of the deckhouse facing the blast. You've got to consider snow and ice loading on the deckhouse and topsides structure, Hoffman finishes, not only for the stability of the ship but for the thickness of the steel which has to support it.

Returning to the beam analogy, imagine that the ship is one huge, long I-beam. The top and bottom of the beam, called the *flanges,* are joined by a vertical member, the *web*. On our ship, thus, the bottom flange would be at the level of the bottom, and the upper flange would likely be the main deck level. Everything in between can be thought of as constituting the web.

Considering the orientation of the kinds of forces making the ship want to bend, sag, or hog—gravity, the usual movement of a ship at some angle to oncoming waves, the fact that the ship is much longer than it is wide—it turns out that the upper and lower flanges are where most of the bending takes place. The ship does not want to bend left or right so much as up and down. Therefore, it is at the level of the main deck and keel that one would expect to find the thickest steel. All of the decks in between, though they too must bend, do so far less in absolute terms than the top and bottom flanges, and so use thinner steel.

In general, most hull plate is one-half inch or five-sixteenths inches thick. The areas of greater stress along the main deck of the ship and along the keel, where the most bending occurs, has plating up to three-fourths of an inch thick. Critical spaces— engine rooms and VLS, for instance—will have additional ballistic plating, even sections of double hull, to contain or diminish the blast of any random incoming projectile that pierces the hull, or at least focus the explosion outward, away from the critical space.

In practice, the thickness of the hull plate and the size of attached reinforcing stiffeners in different areas of hull, deck, and bulkhead vary enormously and for many reasons. The very radical flare of the bow—the part that hits the waves first—has both thicker shell plating outside and very closely spaced, heavier-gauge web frames inside to take the slamming loads when the bow crashes down into a wave.

At various points below the waterline, *sea chests* are built into the hull. A sea chest is a box built into the inboard side of the hull with a strainer on the outside and pipes on the inside to draw sea-water into shipboard systems for cooling and desalinization. Anywhere you make a hole in a steel structure you have to compensate, so the plating around the sea chest will be a little thicker.

Inside the ship, beams support the various decks, and through the beams must run all of the distributing pipe, vent, and cable systems that carry all kinds of things—fresh water, cooling water, halon gas, jet fuel, heated oil, oily waste, diesel fuel, sewage, electricity, air both hot and cold, even information—from one end of the ship to the other.

One of the most common sights at the yard, whether in Hardings, the Assembly Building, or aboard any unfinished ship on the ways or in the water, is a welder with his head stuck into the corner of some assembly, foundation, or ceiling where one thing—length of pipe, cable, section of duct, even another stiffener—penetrates another thing—web frame usually, also bulkhead and deckplate—at any angle. This penetration always requires additional support around the hole, whether it's a flat oval collar welded to the inside of a circular opening in a beam, or one of the tens of thousands of tabs of steel just a couple of inches on a side welded across the narrow gaps created when a stiffener is laid into or goes through a web frame.

One of the practices where shipbuilding departs from other heavy industrial enterprise is in what happens after a ship is "designed." With the DDG-51s, the Navy in essence said let there be an engine room about here, with this engine and this *reduction gear* (a gearbox and clutch that steps down the faster turbine rpm to the slower prop-shaft rpm) and all of the associated intakes and uptakes (exhaust), electronics, fuel sources, and so on.

The initial BIW design team preparing the bid further refined this, with the help of other engineering groups, through three stages of functional, transition, and detail design into fairly precise overall descriptions of, for example, every hull plate, their thicknesses and where they are attached to each other (called a shell expansion), and various drawings that describe the ship deck by deck (the *scantling* plan), within each deck space by space, and within each space item by item.

None of this, however, is really useful to the deckplate mechanic—the burner touching electrode tip to steel to cut a plate, the welder attaching T-beam stiffeners to a section of hull. Before any mechanic with a tool in one hand even approaches a piece of steel, you can bet he's consulted the piece of paper in his other hand or taped to the plate—a work order that has its origins in the Engineering Mold Loft.

Craig Whitman, a BIW project engineer, describes the work of the Mold Loft thus: We start from the plans and translate that into data which is used to create the structural parts that go into the ship—structural parts as opposed to outfitting parts. The structure is the steel in the hull, the T-bars that frame up the hull, the bulkheads, the foundations which go under pieces of equipment. We have pieces of equipment ranging from the very large engines down to little electrical boxes. Just a tape recorder, for example, would have to be screwed into something to keep it from sliding all over the ship. What it gets screwed into is a foundation.

In the case of a plate part, Whitman continues, we take a look at the drawing and find out exactly from the model what the size of that piece is going to be. What the actual commands will be to drive the burning machine. To burn it out of the plate. Where it is going to be assembled. What marks have to be put on the whole plate to indicate where other pieces are going to be assembled. The orientation, because it's going to assemble in a certain configuration, and not necessarily the configuration that it's going to have in the ship. Some things are assembled upside down or lying on their sides; we try to assemble in a position where the welding can be done in the downhand position [where the welder does not have to raise his arms above his head] as much as possible. It's easier, cheaper, faster.

Craig Whitman and his colleagues go through this complex descriptive process for each one of the approximately 150 thousand different *types* of structural pieces that go into one ship. Though these days the calculations can be performed on a Computer Aided Design system that allows the designer to construct and manipulate figures in three dimensions, as recently as the 1950s the Mold Loft at BIW was a large, flat, open space on the third floor up at the Hardings plant where you could find, scratched into the freshly painted floor, full-size, the complete lines of the ship.

Even though today the computers do the number-crunching, the principles of the Mold Loft and its underlying purpose haven't changed. While the naval architects, engineers, and designers may have more or less accurately described what the ship is going to look like, it is the Mold Loft's job, in a sense, to break that master plan down into manufacturable pieces, to specify exactly what the pieces are and how they are to be cut, fitted, welded, and eventually joined into a whole.

While there is no need to get lost in the intricacies of the process, one of the Mold Loft's most difficult tasks is to define exactly where everything is on a ship in relation to everything else. Unlike an architect's plan for even the most intricate of houses, the ship's exterior walls, so to speak, can change in three axes from foot to foot. Even in a house with curved walls, the curves are usually uniform in some respect—of a given height, flat in the vertical plane, of an unvarying thickness.

Absolutely none of these holds true for a ship's hull, which tapers to nothing at the bow, fills out amidships, and can change again in the stern. Stand underneath a DDG-51, and the first thing you notice is that the shape of the ship in the keel area is uniformly curved only in a very short stretch after the massive upthrust of the bow and before the abrupt eruption of the prop shafts about two-thirds of the way to the stern.

In order to adequately describe and locate any particular place on these ships—and thus build it!—you first have to be able to define it in three dimensions.

U.S. Navy ships are delineated from bow to stern in one-foot *frames*. In the DDGs, Frame 1 is located exactly one foot from the stem—the foremost point of the bow. Frame 506 is 506 feet from the bow and thus at the very stern. The Mold Loft generates a series of plans, one of which is the *body plan,* giving a vertical cross-sectional snapshot, a slice of the ship at any given frame line. There is the *lines plan,* which describes what the dimensions would be for any waterline—a horizontal slice from end to end at different levels from zero, the very bottom of the keel, then two feet up from the keel, four feet up, six feet up, and so on. In addition, locations are defined by *feet off centerline,* the centerline slicing the ship neatly in two equal halves longitudinally.

In practice, says Craig Whitman, the way we actually do the job is, we do a computerized model. After we create a lines plan from one end to the other, we then do models which describe a unit. Somebody has decided our Unit 2130 is going to run from here to here, and it's going to be from this deck to that deck. So we model every piece of the structure, and from this we generate the pieces. For each piece we create an AutoCAD drawing of the piece in the position it's going to be burned—looking at it flat.

Every mechanic *on the tools*—doing something other than pushing a broom—uses the work of the Mold Loft every single day. If the welder has to join four pieces of steel, the work order not only describes each piece (thickness, size, material, and so

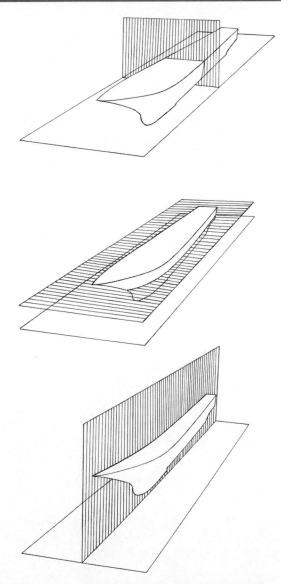

Top: frame line; middle: waterline; bottom: centerline.

on), but gives a welding plan guiding the order of welds, directs the type of welding machine and flux to be used, shows how beveled edges or edges of unequal thickness are to be oriented, and always includes a drawing of the final product as well as a record of where it's been, where it's going, and every process that has been inflicted on it along the way.

■　■　■

When Whitman talks of "units," he is actually touching on one of the more unique aspects of how work is done at the yard. While he is oriented towards the micro view, the work of the yard is not. It can't be, for even the most dedicated systems engineers have not been able to wrestle naval shipbuilding into the twentieth century in terms of reorganizing the work into anything resembling a repetitive assembly line.

In essence, the yard is putting an enormous jigsaw puzzle together. For efficiency, however, instead of starting at a corner and working in, they start dozens of teams working on smaller portions simultaneously, then assemble the portions one at a time. While not a hard-and-fast (and certainly not exhaustively complete) picture, the following is a reasonably accurate breakdown of what is done where in the various facilities that make up the yard:

The Hardings Fabricating Facility shapes hull plates, burns all the steel and other materials to size, welds up many of the small and most of the larger foundations, makes all of the air- and watertight compartment doors, paints everything with primer and finished pieces with epoxy, and prepares most of the flat bar and angle bar, the T- and I-beams for web framing and stiffeners. The largest structure it produces is limited to what will fit on, and can be conveyed by, a tractor trailer.

The East Brunswick Manufacturing Facility also makes assemblies and subassemblies, but these are in the more specialized areas of pipe and ductwork. They manufacture the vast majority of pipe and duct runs in pieces, welding them up or connecting them as applies, for shipping to the yard.

AB, the long green Assembly Building, takes in products—plate, shaped steel, foundations large and small, pipe and duct runs—at one end, forming them up into larger configurations called units. Imagine the ship cut into forty-eight-foot lengths, cross-sections, then each cross-section divided again horizontally into four more sections.

The ship is divided into about seventy of these assembly units, and they tend to be large—again, each represents a slice of the ship about fifty feet wide, forty-eight feet long, and usually one

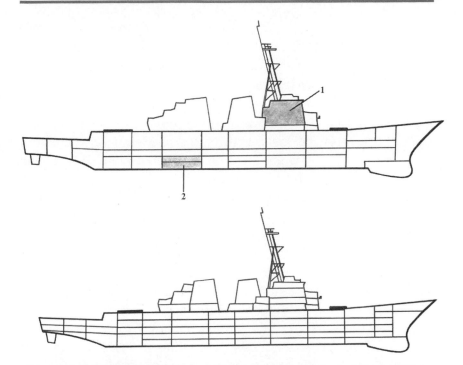

Top: The forty erection units that make up the ship, each of which is laid in place on the ways by the crane. (1) is Unit 4240, the deckhouse discussed in Chapter 5. (2) is the aft main engine room unit, which both appears in this chapter and is seen in the photo in Chapter 6, being erected. Bottom: The seventy-plus assembly units that are fitted up mostly in AB and joined together to make the erection units above before going out to the ways.

deck in height. As each unit moves north through the building, it gets more and more complete and is then welded to its mate above or below to become an erection unit, of which there are forty. When it goes out the door headed for the ways, the unit has gone from duckling to swan. It's still about fifty feet long, but can be thirty feet tall and up to three decks in height.

The reality is that some of the units will barely fit out of the doors of AB, and, weighing more than 300 tons, are approaching the safe combined lifting capacity of the two way-side cranes, which must hoist them off of the steel pylons that support them on the pavement as they undergo final adjustments and lower them gently onto wooden blocking on the ways to be joined to their neighbors.

After AB, the ways are probably the most impressive feature of the yard even though they are usually invisible, lost under the ships taking shape above them. "Ways" are nothing more than a ramp sloping slightly from shore to water, a means of launching ships that has been around since the Phoenicians. People at BIW use the word "ways" generically, but there are more precisely two parts to the ways. The *launching ways* appear as two wide elevated parallel tracks, with a sunken space between them known as the *ground ways,* like the grease pit in a garage. The *building ways* are the open spaces to left and right of the raised launching ways, and they are dominated by the seventy feet of scaffolding that envelops the ship on each side, providing access to the ship at every level for the ceaseless work on hull and interior. The sunken space between the launching ways is also informally called the building ways.

As each unit is completed, it is erected, set in place with the lowest (and flattest) point of the keel resting on foot-square centerline blocks. Where the hull begins to curve up, short stubby *shores* are inserted at regular intervals, kept from kicking out by steel clips welded to the hull. Farther up the side of the hull, where it begins to go almost vertical, angled side shores, likewise secured with clips, are butted against it on port and starboard. Centerline blocks, inner shores, and side shores are together known as *shore support,* and they remain in place during the entire erection phase of construction.

The yard lays the thirteen keel units first, beginning amidships and working fore and aft piece by piece. Where the hull loses depth abruptly at the fantail over the props, going from thirty-six feet to sixteen feet between deck and keel line, each of the two units that make up the fantail is but a single piece, from deck to keel. Almost the entire main deckhouse, which houses the bridge, is also assembled and erected as one unit.

Several months before launch, the launching ways are covered with a slipcoat of grease and wax, and beds of *sliders* are laid down, foot-square timbers yoked together to make a four-foot-wide bed running the length of the hull on each way. Built up from the sliders to the hull is an ingenious network of wooden blocking (called *solid packing*) whose top pieces are molded to fit the curve of any part of the hull. Tying the blocking together from port to starboard, and preventing the weight of the ship from kicking it out, are tie-rods and cable. (See Chapter 8.)

In the hours running up to launch, the shore support is dismantled, and the weight of the ship comes to rest on this newly created *cradle*. At launch, ship and cradle slide together down the launching ways into the water to a point where their combined buoyancy floats the ship off.

Though the looming AB and the inherent drama of the ways defines the yard, there is much more going on here than just assembly. Blast and Paint is a dark, hulking, airtight building where whole units have their skins bombarded with steel shot down to bright metal before receiving coats of dull red primer and haze-gray marine epoxy. There is the Machine Shop, home to a group of mechanics with the special skills needed to prepare the rudder assemblies and uncrate, tune, and test the sections of prop shaft.

Everywhere in between these large, open spaces are crammed the masses of engineers, designers, and managers whose work on paper supports the shipbuilding process, directs and guides it, and to some eyes what is most important, documents it according to U.S. government and naval contracting regulations such that the yard gets paid for it.

Over the course of constructing the DDGs (and the frigates and cruisers they built for the Navy earlier), the yard has pushed and pushed to build larger and larger units, to erect each unit in a more advanced state of completion, and to have each successive launch represent a ship closer to sailaway in both absolute weight and level of overall completion. It used to be that units were, comparatively, mere steel shells at erection, with some foundations installed, some pipe and duct in place.

Things have changed from those days. Walk into the Pre-Outfit 2 (PO-2) building on the river, and it's like a scene from *The Third Man,* where the camera pans down a street and the audience looks into houses whose facades have been ripped away by bomb blast, revealing floor by floor the complete-down-to-the-lampshades living rooms and dining rooms of the people who lived there. Except in PO-2, you see an erection unit on end, propped up on massive steel pylons, open to the eyes from floor to ceiling. Instead of living rooms, there may be machinery or electronics spaces, but they are so complete that it's hard to convince yourself the unit in front of you hasn't been cut, perhaps, from a finished ship.

Deck by deck, all of the lights shine overhead, revealing entire compartments outfitted with not only their foundations but also the equipment that sits on them. Bulkheads glow with their finish coat of paint, airtight doors swing soundlessly from their hinges, and ladders and hatches are in place and functional. From a halon gas firefighting system hangs an inspection tag showing that the seals and pressure have already been checked, while banks of communications equipment, also installed and tested, have been boxed over in plywood to protect them until the unit is joined to the rest of the ship.

While it is certainly a sensible systems-engineering goal to finish as much of the work as possible before the ship is actually launched (some things can only be done with the ship in the water), the more pressing reason has much more to do with rivals nipping at BIW's heels, with the modern era, with global competition and political wrangling; really much more to do with the forced modernization that must go forward successfully for the yard to continue to exist.

I think, Russ Hoffman remarks wryly, that to the uninitiated it looks as though, if you can build a ship, you can build any kind of ship. That's certainly true, he continues, his voice at odds with the words, but there are costs and competitions. The question is, why can't we build an oil tanker like Mitsui [a major Japanese commercial yard] or aircraft carriers like Newport News or the Koreans? The answer to that question has everything to do with what the product is. The characteristics of naval combatants are that they have relatively light scantlings, the shell plating is thin, the stiffeners are smaller, the emphasis is on performance—speed—so the weight needs to be kept down.

And the complexity, he says, the complexity that is squeezed into a package—a ship—whose size is often driven by *political* considerations more than technical ... Yes, BIW has become extremely good at putting a lot of piping, and ventilation and electrical cable and all the equipment associated with those three systems, into confined spaces, and doing it in the production process such that it's efficient.

What I'm talking about, he says, is pre-outfitting all of those systems [with the unit] on blocks before those units are erected on the launching ways. And, to a large extent, testing it before those units are erected so that there's less work to do after erection. The best comparison is that, for something that takes us an hour to do in AB, it takes us five hours to do on the ship in the water. It's five times more expensive. That makes sense. Because when a mechanic is working in the shop, it's enclosed, his tools are there, his stock is there. When he's out on the ship, he's got to walk all the way down there, kind of wend his way down to where he's working, to get his stuff down there, so it's logical.

That's what BIW is good at—and even good at it to the admiration of foreign commercial shipbuilders, Hoffman finishes. When we were involved with Mitsui in this technology transfer program, the Japanese were scoffing when they came into the yard for the first time. An American shipyard, what could they possibly do here? When they saw the extent of pre-outfitted equipment and systems on our blocks for the DDGs, they were impressed. They wanted to know how we did that.

Others add that, if working on the water is five times more expensive, working on ways isn't an ideal situation either, costing approximately three times as much as inside work. This is due partly to the compensations that have to be built into every calculation because you are building on an incline. To erect a bulkhead wall at right angles to the deck out on the ways, the shipfitter must use a level with a shim compensating for the few degrees of angle from shore to water. Not only does the yard have only three functional sets of ways, when it could use four, but some of the ways are canted at slightly different angles, meaning that tools must be specific, depending on which ways is in use.

As impressive as it is, even to the Japanese, pre-outfitting and all the Yankee ingenuity in the world have not been able to make up for a few unavoidable facts that are fundamental to the way in which the yard has built and launched ships for more than one hundred years.

BIW is one of only two shipyards in the United States to erect and launch ships of such size down ways into the sea. To put it bluntly, the yard can pre-outfit until it's blue in the face, but will still, one day, run up against the maximum weight limitations of the ways. The ever larger, ever more complete, ever heavier units are a good plan only if there is another method of launching the ships.

That is indeed the shape of the future, a $210 million ten-acre level platform with, on the river edge, a detachable submersible drydock. Ships are erected on the platform and towed on rails out to the drydock. The drydock is floated out into the river and flooded, and the ship eases free.

With the passing of the ways, there is no question that the yard will lose a lot of history. The first drydock launch will bring to an end a tradition going back to the very roots of this place, this town on the river where shipbuilding first began in America. The old-timers know it, feel it, even reminisce openly when they're standing around on launch day. The younger mechanics seem to understand, however, that the change is what will allow them to continue to have work as shipbuilders until they, too, retire, ceding their places to the children they now dandle on their knees.

BATH, CRADLE OF SHIPS

Many communities in the U.S. are historic shipbuilding sites, but most of them no longer build ships. You can think of it as a historic thing, as part of a tradition, but for us it's still happening. This isn't 75 percent scale like Disneyland. This is real.

— James Upham, Bath City Planner

The city of Bath, Maine, is unlikely to win any awards as the prettiest place in America. Shipyards are generally not pretty, and the Bath Iron Works shipyard, which takes up fully half of the city's riverfront, is no exception. The yard's two tallest cranes, both of which reach their arms up more than two hundred feet into the sky, can be spectacularly beautiful wreathed in fog or dappled with snow, man-made storks stalking the horizon. The five-hundred-foot-long Arleigh Burke–class guided missile destroyers that BIW builds, three of which, nearly complete, usually rest quietly at the piers in front of the yard, have a beauty more chill than thrill, sleekly menacing in the haze-gray that cloaks their rakish lines.

But behind, between, around, all about is the yard, ten acres of open riverfront strewn with the detritus of industry: pieces of ships, and pieces of pieces of ships; piles of rusting scrap, pipe, plate, and beam seemingly abandoned outside hulking factory spaces whose gaping doors leak sparks and fire and the reek of burning steel and marine epoxy. The air is heavy with the weight of noise as well, the various shrieking sirens, bells, and whistles of moving machinery, the PA system blaring messages, the ringing boom of steel being shifted, the teeth-jarring clang of a shipfitter's sledgehammer.

If they gave an award for sheer spunk, however, for grit and energy and perseverance through good times and bad, Bath would surely win. It is a place that has formed its character through four hundred years of building ships of wood and steel, where the blast of the yard whistle punctuates births, deaths, and marriages, and where the lunch pail is more respected than the briefcase.

Nestled comfortably on a small rise above the Kennebec River, its graceful, aging houses in every style from early Federal to Victorian flowing down the hillside to the wide, slate-gray river, Bath has a past and present as the "City of Ships" that is never far from sight or mind. If the verdigris-covered weather vane atop City Hall—capturing the four-masted barque, *Roanoke,* built and launched by Bath's Sewall shipyard in the late 1880s—represents the end of the city's dominance in wooden shipbuilding, the name of the Bath Middle School football team—the Destroyers—surely signals its primacy in our own times.

Downtown Bath is split awkwardly by the raised viaduct carrying Route 1 over the Carlton Bridge. South of the bridge, the shipyard takes up all of the riverfront, penetrating back several blocks before residential rental housing takes over. North of the viaduct lies the town proper. Fronting the river is Commercial Street, a sparse stretch with underutilized businesses, a public landing where visiting boats can tie up free of charge, a park of winding paths, benches, and picnic tables, and another large lot currently used as a staging area for prefabbing sections of the new bridge soon to replace the aging Carlton.

Stand anywhere on Front Street, the next block back from Commercial and the heart of the downtown, and it's as if you've been transported into Frank Capra's proto-America. Three- and four-story brick buildings line the street, with merchandise spilling out of ground-floor storefronts and one-way traffic cruising slowly past, wide redbrick sidewalks with black-painted gaslights. Even the No Parking signs are wooden and hand-painted cream with green accents. The street is filled with the good-mornings of neighbors going about a day's business: a haircut at Ben's Barber Shop, filling a prescription at Wilson's Drug Store, cashing a check at Bath Savings Institution, sorting through the bins of specials at Reny's discount department store. At the far end of the street, tourists wander from antique store to coffee shop to craft shop.

Wherever you are in Bath, the cranes dominate the skyline. This picture was taken from downtown Front Street.

Front Street rises gently along its length, its highest point at the southern end, closest to the bridge and shipyard beyond. This low crown is occupied, appropriately enough, by the Davenport Memorial City Hall, an imposing three-story limestone edifice with a semicircular front bay, monumental columns, and a cupola holding a bell cast in Paul Revere's foundry. Inside, the visitor discovers a place of antediluvian charm—warm brown linoleum floors, potted plants, spacious benches, and wooden doors with inset pebbled glass, everywhere the bright gleam of varnish, and an antique elevator with accordian brass gates.

Anywhere else the 1920s elevator would have been ripped out as an insurance risk and the floor covered over with industrial carpeting, the oak and pine doors replaced with steel, the wood trim and signs exchanged for easy-wearing laminate and plastic. Here this relic of the past not only has been preserved but is in loving everyday use, down to the "Last One Out? Please Check The . . ." list taped to the inside of the heavy front doors.

The first-floor walls are hung with the comforting icons of small town life, the local police officers, their faces smiling out from baseball cards, photos of the police chief crouching next to a baby stroller, the community bulletin board hung with a "Rabies Alert!" and the Bath Municipal Band evening performance schedule. An offer of low-cost fixer-upper loans from the housing authority competes with notices for Victorian house tours, a coming increase in water rates, the opening hours of the Bath History Museum run by local middle school students.

A brass plaque dominates one stretch of wall, the severe countenance of Sumner Sewall in bas relief over a rather lengthy paean to this local son, state governor during the difficult years of the second world war, member of a "famous family of shipbuilders . . . a colorful, vigorous, friendly, and effective leader." At every step you take in this town, the grand old names of the city's shipbuilders and shipping magnates hit you in the face: Sewall, Morse, Coombs, Newell, Patten, Hyde, Drummond, Donnell, and so many others.

Almost nothing is known about the area's first settlers, the members of the unsuccessful 1607 Popham Colony just down the Kennebec from Bath. What they did leave, however, was a legacy, for some of the survivors sailed home to England in the pinnace,

Virginie Sagadahoc. As the official publication of the Sesqui-centennial of Bath, Maine, notes, it was "the first sea-going vessel built in North America by Englishmen. This event allows Bath to claim to be the 'Birthplace of American Shipbuilding.'"

Forty years before the town was first legally constituted in 1781, one Jonathan Philbrook began building two ships in town. The number of yards and vessels launched grew slowly after the end of the American Revolution, until by the 1850s Bath had blossomed into one of the ten largest commercial shipping ports in the country, its waterfront a teeming warren of launching ways and chandleries, sail lofts and workshops, the expanse of water north of what is today the Carlton Bridge having become known locally as "Schooner Way" for the forest of masts rising above its waters.

The great wealth that flowed into Bath, taking the form of Washington Street's sprawling Federal homes and imposing brick and stone public buildings downtown, did not come from build-ing ships directly. At that time most shipbuilders built ships not to sell to others, but to use to trade. Bath had its share of shipping magnates, men like Arthur Sewall, who in the latter half of the nineteenth century controlled one of the largest merchant ship-ping fleets in the country. Such men brought prosperity and jobs, certainly, but also new ideas and influences from abroad. Some of the more unusual—and enduring—symbols of this foreign expo-sure are the exotic trees the sea captains brought back and planted in their spacious yards, cork oak from Asia, South American cucumber magnolias, Chinese Kwanzan cherry and ginkgo trees, dramatic bloodleaf maple and palm heart nut trees from Japan, elegant lindens and Camperdown elms from Europe.

In 1884 an enterprising young man named Thomas W. Hyde, who had come home from the Civil War a general, incorporated the Bath Iron Works, Ltd. Seven years later, having successfully transformed itself from a maker of auxiliary shipboard equipment and deck machinery like windlasses, power capstans, steam boil-ers, and engines into a full-fledged shipyard, BIW launched the 205-foot *Machias*, a U.S. Navy gunboat rigged for sail and steam, and the first ship whose keel the yard had laid. By the turn of the century the ways were kept busy as the yard launched luxury steam yachts for rich Bostonians and New Yorkers, passenger steamers, torpedo boats and gunboats for the Navy, and lightships for the U.S. Lighthouse Service.

With World War I, the yard entered a boom and bust cycle from which it has never really been able to extricate itself, expanding rapidly during war years with a subsequent collapse soon after. During the peak years of 1915 to 1920, for example, BIW launched twenty-two ships representing a total of almost 25,000 tons of shipping. Sixteen of those ships went to the Navy, and fifteen of them were 314-foot-long 1,000-ton destroyers with which BIW was already making a name for itself.

From 1921 to 1925, though thirty-one hulls were launched, the total tonnage was only thirteen thousand. A closer reading of BIW's *List of Ships*, a concise summary of all hulls and their briefest specifications, however, reveals that the nature of the yard's business had changed markedly. Not a single ship was a Navy contract, and sixteen of those thirty-one hulls were 20-ton wooden schooner yachts on which the yard did not make large profits. The *List of Ships'* entry for Hull 97, almost pathetic in its brevity, reads, "ROMANY, 6-Meter Wood Sail Yacht for Frank Paine. 23'9" long, 6'6" beam, 5'5 ½" draft. Delivered July 1, 1924." By then the yard was reduced to building turbine sets as a sub-contractor, back to the point where it had begun thirty years earlier. As Ralph Snow notes in his fine encyclopedic history, *Bath Iron Works: The First 100 Years*, "No job became too small for the shipyard that had once built the fastest ships in the US Navy, whether it involved repairing and retinning milk and ice cream cans or analyzing samples of suspected moonshine for the County Sheriff's office."

Caught in the downward spiral of terrible postwar business conditions, and with a floundering management and onerous interest payments that eroded its capital to the point where it could no longer satisfy the bonding requirements of large ship-building contracts, the yard went slowly, quietly bankrupt. It was bought at auction by a bottom feeder from New York named Friedberg, who in just a few months had stripped out and sold off all of its tools and machinery, with the leftover stock let go for scrap. There would be no launches in 1926, or in 1927.

The moribund yard changed hands several times over the

next two years as scheme after scheme to revive some sort of gainful enterprise in the vast buildings fell apart. Finally a former BIW engineer, William "Pete" Newell, managed to lease the yard, and with the backing of a few prescient individuals and the town fathers, outfitted it with the basic machinery and tools necessary to get back to the business of shipbuilding. Over the next few years he slowly resurrected the yard, initially on the strength of high-quality yachts like the *Corsair,* the three-hundred-foot floating palace it built and launched for J. P. Morgan in 1930.

The relationship between city and shipyard has evolved over the last hundred years from the typical nineteenth-century one-sided, patriarchal, yard-gets-what-yard-wants into more of what a psychologist might characterize as a modern codependent marriage, so closely has the fate of the two become intertwined in recent years.

The dependencies are on both sides, of course. Today 7,800 people work at the yard, of which about a third are estimated to live in Bath itself. The yard has been, since the postwar era and in some years earlier, the largest private employer, not just in Bath, but in the entire state of Maine. As its business has grown, so has its appetite for surrounding land, as the yard has gobbled up neighboring shipyards and businesses, even moved across Washington Street, demolishing and relocating homes as more and more of the construction shifted from ways and water to inside ever larger buildings.

The company is the city's single largest source of property taxes, its contributions generally making up from 20 to 30 percent of total revenues, depending on the year. A good portion of its $300 million annual payroll goes home in the pay packets of workers who live in and around Bath, with another $23 million paid out to suppliers, some of which are also in Bath.

But the residents of Bath pay a price, too. People move to Bath for jobs at BIW, as a result of which, says Jim Upham, city planner, this is the most densely settled part of Maine north of Munjoy Hill in Portland. The city is only about ten square miles, versus fifty square miles for most Maine towns. It's a port city, it's small, it's surrounded by the river and other geographic constraints, with small lots, a heavily built-up urban area.

Upham is tall and spare, partial in dress to woven woolen ties, simple shirts, and khakis, which fit right in with a generous gray beard and mustache. Under brown hair going salt-and-pepper, clear blue eyes shine out of the kind of middle-aged face that won't change much for twenty years. He is thoughtful of mien and slow to speak, likely a quiet voice of reason amidst the political wrangling of city council meetings.

Bath is an *urban* area, he continues. I can leave City Hall and go walk and have lunch someplace; I don't have to get in my car and drive somewhere. It's just very compact, and in ways very pedestrian-oriented. It's not right on the coast, but it's almost a coastal town. My sailboat's only fifteen minutes away. And it's a close-knit community. After being here for four or five months, I was made to feel part of the community, I was involved in things. I *live* here.

Bath still has a downtown, a functioning downtown, Upham says emphatically, right adjacent to the very heavy industry, and they function relatively well together. We have people come and say it's great to be here because this is real. They can be here, they can see the water, they can walk Washington Street and see the shipyard.

The downside of all that, Upham says with a professional frown, is that we have seven thousand people who commute into our city every day. Of all the places I've worked in Maine, I've never seen parking such an overwhelming issue as it is here in Bath. If someone parks here for more than two hours, they're going to get a ticket. That's because, if you're not on top of that situation all the time, the workers creep out into other parts of the city with their cars . . . just a fact of life.

No one is more aware of Bath's advantages than the city manager, an appointed official responsible to the city council who runs the city affairs from day to day. Sitting in a large oval meeting room, an airy space of old photographs and prints of ships and shipping, rich woodwork, and many windows looking out over Front Street, John Bubier radiates confidence and good cheer. As befits a city booster, he sports a maroon polo shirt with the city's symbol, a square-rigged sailing ship, picked out in white where the polo player would be.

As he speaks, his sunburnt features animated by a crooked smile under trim gray hair, his words are forceful, delivered with a confidence and overt enthusiasm unusual in this place of closemouthed workingmen—deckplate mechanics, they're called here—descendants of generations of laconic fishermen. It's easy to be quite taken with him, for it soon becomes clear that this is a man with a plan.

Bubier has reason to be happy, for a state court has recently thrown out a taxpayer challenge to the city's agreement to divert $81 million in BIW property taxes over twenty years to help fund the upcoming $300 million modernization of the shipyard. He's happy, too, with the Navy's recent award of $2 billion worth of DDG-51 work, part of an ongoing authorization package which virtually assures BIW a continuing role in the richest, most significant shipbuilding program of this century, a virtual guarantee of at least a decade of future work. Further, the yard has advanced in the bidding for SC-21, the Navy's Surface Combatant 21st Century program and consequently next century's biggest plum.

Yet the town's problems—John Bubier's problems—haven't really changed from the far bleaker time a year ago when he first took over the job. The yard's future was anything but certain, there were rumors of coming labor unrest, and there was some small amount of rancor among inhabitants left over from a successful BIW suit to recover more than $2 million in improperly levied taxes.

Over the past year the city, with the help of development consultants, has been trying to redefine itself, to differentiate itself from the yard both symbolically and economically. As city planner Jim Upham points out, We've got all of our eggs in that one basket from a tax base point of view and from an employment point of view. We'd love to have this shipyard here long-term, but we believe that we have to not so much get our eggs out of that basket, but get more baskets.

The danger is more the attitude that the community develops around itself, Bubier says. It's kind of a *We don't have to worry because we have the yard*—and therefore we don't have to do anything. We don't have any major concerns economically outside of it. It's kind of lulling yourself into a sense of wellness.

He has deeper concerns, too. Why has Bath, he wonders aloud, not taken its place of prominence? Was it simply the location next to Brunswick? Was it simply the way it developed over

time with the Route 1 viaduct? One of the things we found interesting and isolating is the comment, *Gee, I never knew there was
a city there.* We know that there is this place, that when you cross
the bridge and look back, you see the ships. But you can't get
there from here! Coming up from south, the first time that you
know there is a downtown currently is just after you made the
decision to cross the bridge. Unless you are really brilliant and a
really good driver, you end up on the bridge and can't get back!

Hence the plan, not as yet finished, but about which Bubier
can be very convincing, especially when his blue eyes get to flashing and his voice falls into the cadence of the revival tent. It is a
plan, he says, to look at the gateway to the city, the business dis-
trict, and the yard, and how we build an integrated approach. Part
of that will be continuing on with the gaslights, those parks in the
downtown area, developing green spaces where we've now got
nothing but macadam, building a rail and sail facility.

Here his speech quickens, his finger flashing from point to
point over an aerial map of the city and yard, Route 1 splitting
them neatly in two. For example, he continues, one of the initial
pieces is, you've got your functioning rail line here, you've got the
third-highest-traveled road in the state of Maine here, and you've
got a major river that now can be travelled to Augusta and beyond
. . . we can put you in Boothbay Harbor quicker from right here
than you can drive! You could stay at a hotel here that's part of a
package, take a shuttle down into the harbor and spend the
evening down there, have dinner . . .

But he's a realist, too. I know, Bubier continues, that the overall majority of people who live here support this shipyard and
trust in the yard. I also know there is a certain amount of give-
and-take. The town doesn't always get its way and the shipyard
doesn't always get its way. There is this kind of love/hate thing
over the years. We love them because they provide a tax base; we
don't like them because they take up our parking. We love them
because they provide a tax base; we don't like them because they
tell us what to do. I mean, it's a conflict.

In the mid-'80s there was a *major* battle over tax evaluation,
Bubier observes. This got people cranked. The fact of the matter
is that the yard was right and we were wrong. And we wound up

paying a significant amount of money to recover from that. But even in the face of that, last year when push came to shove and the issue was, would you put up $81 million bucks for this organization, the people said yes.

Back on Front Street, blue sky above and the sun throwing shadows, the street crowded with people off to lunch, the city feels alive, prosperous, just another place where the '90s economic miracle is alive and well. Raise your head, however, and a far more fragile reality looms. Everywhere in Bath—by the river or up on the hill, from nearly every window in every building with any kind of view south, the two cranes parade across the sky, their booms moving in an achingly slow ballet.

What would all these people do if the shipyard, a billion-dollar company, suffered a long-term slowdown in new orders, or closed altogether? If its corporate parent, General Dynamics, announced one day that it was moving all the work south to the shipyards it owns in Rhode Island or Connecticut? Thirty years ago there were twenty active naval shipyards in America, and today there are six. What would the people of Bath do?

Unfortunately, Jim Upham says a little uncomfortably, there hasn't been any planning around that. I don't know if it's something everybody's afraid to talk about, to think about. . . . John Bubier and I, we're creating more opportunities here so that other things can happen, things that can stand on their own. We'd love to have that shipyard here forever, but we're diversifying.

Tom Hoerth and Mary Ellen Bell moved to Bath in 1987, when they got married, living first in an apartment, then buying a house on Middle Street just one long block behind and above the yard. Theirs is, like many in Bath, a two-story frame house of blue-gray clapboard with white trim, built into the hillside. The gravel drive climbs steeply up the hill, dead-ending in a narrow parking space whose uphill border is an eight-foot retaining wall of railroad ties neatly camouflaged by a luxurious wisteria vine drooping its lavender flowers and silvery foliage over the side.

Their front porch is a pleasant place of comfortable white wicker furniture, flower boxes bright with pansies and sweet-

Tom Hoerth and Mary Ellen Bell-Hoerth with their two children, Carrie and Emily, at home just behind the shipyard with the cranes in the background.

smelling purple and white alyssum. Standing on the porch on a cool, foggy evening in June, you can smell—but not see—the river. The great green wall of the Assembly Building looms up instead, behind it the radar mast of a destroyer abuilding on the launching ways poking its delicate tracery up into the sky. But what inexorably draws the eye, dominating the horizon with its orange and white stripes, is the two-hundred-fifty-foot can-tilevered boom of Crane #15 as it comes slowly about, an unseen burden at the end of its chainfall swinging slowly back and forth. The bright white deckhouse at the base of the boom seems to glow up on its perch a hundred feet off the ground, orange sodium-vapor floods highlighting the great blue "B I W" embla-zoning its side. The wind brings the deep drone of the ventilating fans atop AB, the ululating siren of the crane in motion, seagulls screeching, a motorcycle revving down the street below.

At 8:45 P.M. the yard whistle blows for second-shift lunch. In the living room, a well-used room of Victorian furniture and area rugs, every surface is covered with the books, games, and toys of Tom and Mary Ellen's two precocious girls, Emily, six, and Carrie, nine. The dishwasher chugs away in the background, and small voices float down the stairs as Mary Ellen returns from putting the two girls to bed.

Tom and Mary Ellen are both tall and solid, comfortable in the outdoors, on the water, in the woods. Mary Ellen has perfectly straight strawberry blond hair hanging down to her shoulders, the square-planed beauty of Wyeth's Helga, and a no-nonsense presence perhaps honed over her years of teaching history to teenagers in nearby Wiscasset. She was raised a Quaker, a tradition she continues by going to Quaker meetings and by raising her children with the Quakerly values of simplicity, nonviolence, and meditation.

Everything about Tom is big, hands like dinner plates, handlebar mustache worthy of a Civil War soldier, broad shoulders and muscled arms from working outside as a weekend arborist. A former teacher with a degree in horticulture, the kind of man who decides to renovate a bathroom the way others mow the lawn, he has recently founded an organic produce business, Winter Greens, supplying local health food stores and restaurants with fresh vegetables and salad greens in winter.

Did we think about the kind of town we were moving into? Mary Ellen is saying. Well, no. We felt comfortable. We could afford the house. Having been here ten years, I think we've come to really like this town. Tom's father refers to Bath as a lunch-pail town.

She shrugs off the implied slight—Brunswick with its Bowdoin College rich kids and sumptuous green campus versus Bath's shipyard and working people. My sister lives in Brunswick, she says, we have friends in Brunswick. We don't . . . we both really *like* this town. There's something about it, it's a *real* town. There are working people, I mean, everybody knows somebody who works at the yard, that's a given.

Tom begins counting the neighbors who work there, house-by-surrounding-house, until he runs out of fingers and the two both start laughing. Most do, she says. Some are managers and most not. But yeah, just about every other person at least from High Street on down works at the yard. We've gone back and forth, sell, don't sell, sell, don't sell. . . . But if we did buy another house it would be in town somewhere else.

There's a lot of potential, Mary Ellen adds, it's a huge water-front. It has a lot of character, a real sense of community. I mean, Carrie just spent six weeks with the rec department on an all-girl softball team, a first, ninety girls in town grades one to three. This woman who has run it is just somebody who decided that she wanted to do something for girls. After they finished the season, they had this big potluck dinner for everybody, the Elks Club down on the waterfront offered their space. She gave trophies to every child. There was a whole mixture of people. My obstetrician's daughter was there. There is no pretension. It's real people doing real things, and there's not a lot of social ranking. Tom's on the forestry committee. We really have a strong PTA at Carrie and Emily's school.

Outside, the whistle blast signals the end of second-shift dinner. Tom and Mary Ellen both look briefly towards the yard. There's a whole pattern to the noise, she says. In the winter when you have your windows closed, you don't hear a whole lot. Come spring and summer and fall, you hear a lot, the various machinery, the whistles. We have our bedroom, with the windows open, on the front side of the house. God, it was loud this morning, some big machine . . . it was very loud!

A grinder, it was a grinder, Tom says. You hear the whistles, the loudspeaker. The summer's much worse, his wife continues. Mostly we've tuned it out, that's why we go to Vermont for two weeks in the summer. It does wear on you after a while. The other huge thing is the TV. We don't have cable, we'll never have cable, and we never have clear reception except on Sunday nights.

When they stop moving the crane, Tom says, when they don't move it, you'll find the right spot (he mimes moving the rabbit ears around), you'll go, aaah! You sit down, and you'll hear the *whoooop! whooop!* No! No! Don't move it! and the crane starts moving again. He puts his head in his hands in mock despair.

You hear the union whenever they do their thing, Mary Ellen adds. Their rallies, and the loudspeakers, and the cheering around noontime. You could hear this winter when they were pounding. It was August, her husband says.

The story they tell begins with the single premise that labor relations at BIW are, in a word, bizarre. Nothing captures this more eloquently than the events just before the last contract expired in August. As the union marched from shipyard gate through town one hot afternoon, a middle-aged mechanic named George "The Troll" Trowell collapsed and died of a heart attack. The Troll was a well-respected electrician and also an irascible activist who had proposed such things as five-minute work stoppages and blowing car horns simultaneously to tell the town, hey, we're union and we're proud.

The day after his death, three days before the contract expired, a few mechanics in the AB marked his passing by grabbing hammers and, at noon, beating out a dirge on whatever scrap iron was at hand. Soon, every hour on the hour, thousands of mechanics were taking up their hammers and thrashing away on whatever steel was within reach for five minutes, sending a wall of sound washing over the town. The "Not-So-Silent Protest" was born.

Standing in the middle of the yard, everywhere you looked— into the cavernous AB, on the ships, out in the roadway—you saw the raised arms coming down. The cacophony made people cover their ears and retreat into offices.

Ohhh, that got old fast, says Mary Ellen. They were pounding on every piece of metal in sight, all these different pitches of hammers hitting, *clang! clang! clang!* Really, really loud. Wicked loud, adds Tom. It wasn't a little background noise. It was a pain in the ass. It was the middle of the night, and, I mean, everyone in this area has to get up and go to work tomorrow, why are you busting my chops over this crap? That sucked.

No work got done. The townspeople called the cops when their kids woke up screaming at 3 A.M. The Navy threatened to shut the yard down for destruction of federal property. And all for the Troll, who was surely laughing in his grave, the kind of man, after all, who had been buried in his Harley-Davidson motorcycle jacket.

Tom and Mary Ellen both pause and kind of shake their heads at the unpleasant memory. I know, Mary Ellen continues, that the new drydock has been an issue for a lot of people, neighbors down there. We thought maybe we should look for a place on the other side of town, the north side.

She's referring to the most major part of the coming modernization, BIW's plan to fill in ten acres of river to create a level platform on which ships would be completed, nearly doubling the yard's area and allowing them to launch ships by means of a detachable, submersible drydock that would be part of the platform.

I don't get the impression that we as citizens or neighbors have a lot of control over that, Tom says. His wife nods in agreement, adding, I got the impression this whole drydock, it's a done deal. The other big piece is that people need this industry to stay alive. People need that, and so they are going to make a lot of allowances in order to let that to happen.

Bath is going to be changing so much, Tom says. Between the bridge, the waterfront, BIW expanding. And the library, the new YMCA, Mary Ellen puts in. There is a lot of optimism right now, because that money will come and is supporting the town and the town is able to do those kinds of things. I mean, a new library? We're psyched. A new Y? We're psyched. Those are things that we would use.

The only piece that's not really clicking is the council, Tom observes. It's an old council that's not particularly longsighted. It's who has the biggest wallets. What do you do, Mary Ellen asks rhetorically, with town officials who don't really know what they're doing? Put someone else in, her husband replies. They're petty, they're shortsighted. They don't want to do anything, he says, exasperated. They're stuck in another decade.

The only thing, Mary Ellen interrupts, that impacts our kids' lives is the traffic from 3:30 to 4 P.M. because they're not allowed to walk to their friends' house. There are too many cars and they drive just too fast. Tom has a great description of what it's like when BIW gets out at 3:30. They do not screw around.

There are three stages, Tom begins. The horn blows and the first wave comes. Guys running and jumping, pushing old ladies out of the frickin' way, and they're kicking each other, and the first wave is just flying to their cars and *vroom!* they're out of here. If you're caught in that first wave, they will just run you right over. The next ones are kind of fast-paced, waiting for the first crew to go, waiting in their cars. Five minutes after that, the third wave is like, who cares, don't worry, we'll get out, no hurry. . . . When Mary Ellen was pregnant, I said, please God, don't let your water break at 3:30 because there is no quick way to get to the hospital in the next half hour.

That's the big impact on our life, Mary Ellen says. In terms of what they are building down there . . . my neighbor who works there has two sons a little older than Carrie and Emily. They used to play together all the time. Those boys are into the war machines and the technology, and, you know, I keep thinking, periodically, what are our kids getting from this interaction? This is all about machines that kill, and I'd say that disturbs my Quaker conscience. But, I have a strong belief in what we teach our *own* children. . . .

We've had tremendous congresspeople and senators come out of this state of Maine. From Mitchell to Cohen to Margaret Chase Smith to Muskie—just tremendous leaders. They are very connected to this industry in this state. It's also the economy. Oftentimes, there's a conflict between money versus the people, money versus the higher thing of, do we believe in this industry? And always the money wins out. These are the industries that have supported the Maine economy for so long—the Brunswick Naval Air Station, Bath Iron Works—that we now have this inextricable link between the health of our economy and these industries. We haven't diversified enough.

And there's all these tenement houses, Tom points out. The way the yard shaped the development of the town, and what's gotten developed and what's gotten bought up, what's gotten torn up and what's gotten saved has been I think largely based on their needs.

I remember when I was trying to get a teaching job, Mary Ellen recalls, and went to Morse High School. The principal said something like, we have our professional people, our doctors and lawyers in town, then we have our sort of middle of the road. And then we have our BIW workers! School people on the PTA have talked about how with BIW there has not been a lot of educational aspiration in this town. You know, you screw up and you can work at the yard. Why should you have to graduate from high school? Historically, I think that has happened. We still have a lot of questions about the high school here. We are going to look very carefully about whether this is going to work for our kids.

Even if they end up putting their kids in a different high school, they still plan to stay in Bath. We do, Mary Ellen continues, really see a sense of community, and traditionally the workers at BIW have been really solid people, really good citizens. There's a big cultural blend. Like Washington Street—you've got those bed-and-breakfasts, old captains' houses, and even if you drive all the way up Washington, which is the wealthiest part of town, you still have trailers, little run-down houses. That is part of the appeal, too. There's this big cultural blend.

She begins to talk of the Heritage Day Parade, Bath's three-day July Fourth celebration. They bring in Smokey's Carnival, they bring in great bands and play in the park. They do a dog show. They do a decathlon, two road races, artists in the park. Fireworks. They have this great parade, and we participated for the past several years with our day care and then with the public library. It goes all the way through town and ends up going down through Front Street.

I always end up crying at these things, Mary Ellen says. It must be what little patriotism I have left. There's something about watching the members of your community and knowing people and seeing people, and the Girl Scouts, and the stupid little float that has some homemade thing on it. It's so cool; I always loved that parade, your whole community marching down through town, you feel very connected.

For Mary Ellen and Tom, what started out as a financial decision has become a real choice about a way of life. I feel like we wanted to live in mid-coast Maine, she says, and we ended up in this town because we didn't have any money and this is where we could afford to live. That's the bottom line, and you make your life around that. Who lives in a town where you agree with everything that goes on there? None of us. So I still very much do consider myself a Quaker, and we teach our kids those values. The fact that we have warships being built two blocks away is just one of life's ironies!

HARDINGS

At 6:00 in the morning, with the February sky dropping clumps of wet snow, Bath Road begins to fill with traffic. The vehicles (in Maine, cars and trucks are always "vehicles") come off Route 1 at the Cook's Corner shopping mall, then it's a straight shot past three miles of Wal-Mart and Wendy's, Patriot Lanes bowling alley, a bankrupt lumberyard, the Viking Motel—"Take Time For Fun!" says its sign—through a brown and gray landscape of scrub pine and sand to the muddy parking lot outside BIW's Hardings Fabricating Facility. They come from all over, the mechanics, from Lewiston and Augusta, Pittston, Gardiner, Portland, some driving more than an hour because you don't turn down a job at the yard over drive time.

The lot slowly fills with pickups and 4X4s, the big F-250s and King Cabs with snowmobiles and ice fishing gear in the back, as the start of first shift approaches. It's biting cold and windy as the men straggle up to the main gate, slurping down coffees and sucking in last lungfuls of their morning smokes.

Behind a chain-link fence, the plant runs an eighth of a mile parallel to the road, jogging back at the eastern end to make a rough L shape, four stories of mustard-colored sheet steel with rows of painted-over windows set into the upper floors. Over it all sodium arc lamps cast a sickly light the color of a two-day-old bruise.

"Through These Gates Walk the Best Fabricators in the World," proclaims the sign on the front gate, almost hopefully, it seems at this hour. Worn banners from past workplace campaigns flap in the gloom: "Support ZAP—Zero Accidents Program,"

"Safety is No Accident!" "Cumulative Trauma Disorder—Prevention Through Knowledge." Painted on the side of the building is a bear dressed like a superhero, the BIW crest on his chest and a ballpeen hammer in one paw. "Rework—Can We Bear It?" he asks. The men streaming in the doors don't appear to notice anything, heads down, boots slogging along the slushy roadway.

The whistle blows at 6:30, and a half an hour later the building is awash with noise, smoke, shouts and curses, the sputter and hiss of plasma burners and stick welding, the variable-pitched hum of transformers as the cranes move about on their overhead rails, everywhere the clang, clatter, and boom of steel on steel, on concrete, on the press breaks.

There is very little delicate or elegant about most of the work at Hardings. Here steel is blasted, primed, burned, cut, beveled, welded, bent, sheared, bored, and brazed. It is polished smooth as glass, heated cherry-red, wailed on with sledgehammers, caressed into shape with cool water. The work is elemental, brute-force, dirty, sometimes dangerous. It is a two-story, 500-ton hydraulic press forcing a flat three-and-a-half-inch-thick plate of steel, ram by ram, over a half-round die to make a rudder bearing housing; it is a leather-jacketed welder crouched inside a section of propeller shaft tubing, on his knees, his hood wreathed in sparks and smoke as the great heat of the arc fuses steel to steel three inches from his face.

In one light, Hardings is nothing more than a machine shop, albeit an oversized one. Stock, mostly sheets of steel, arrives by flatbed rail car and tractor trailer at one end, and is always first blasted down to what Navy rules call "white metal" and primed against rust. Now tinted mauve from the weldable primer, it moves to the burning area, which is dominated by four raised tables running roughly east-west, the largest of which is about twenty feet wide and one hundred feet long. Huge hoods loom above the two easternmost tables, their fans sucking up the hazardous fumes of burning steel.

The table by the western wall has no hood and no computer-controlled cutting machine poised above it, as the other three do. Here some plates, usually the thicker ones, acquire a bevel on one or more of their edges to allow the deep penetration of a welding electrode. This is done by means of a small tractor burner that propels itself along the edge to be shaped. Down the deckplate just south, more hand-burning is done in the Shapes Area, black-hatted burners in welding masks using handheld plasma torches to fabricate shapes out of three-dimensional steel, T-beams and I-beams, cutting them to length, notching their ends, burning out slots for the T-bar stiffeners that later will be laid in at right angles.

Hardings burns the largest part of all the stock that goes into every ship, "burning" being deckplate shorthand for cutting. The aging workhorse of the plate-burning floor is a twenty-five-year-old Anderson CM-150 plasma burner. With a high-amp capacity and two burning torches mounted on a gantry arm traveling the length of the table, it handles the largest plates BIW buys—the

A mechanic checks the burning path of the Anderson CM-150 plasma burner amid clouds of smoke rising from the burning steel to be captured, in part, by the hood above.

ten-foot six-inch by forty-eight-foot six-inch full shell plates that sheath the hull, and any thickness of steel up to two inches. Just west of the Anderson is another plasma burner, a Heath, this one using a single torch mostly to burn thinner materials from one-sixteenth up to a full inch.

Both the Heath and the Anderson, because of the nature of the plasma burning process, cut all of the nonferrous materials BIW uses in addition to steel—the aluminum, copper-nickel, and also the stainless steel. Both tables have a lip raised slightly above the level of the flat stock so the table can be flooded with water to cut down on smoke and gas emissions, to reduce the deafening blast of the arc, and to draw off the molten slag left behind when steel is cut.

Plasma burning begins with an arc, the resistance to the passing of an electric current through the tip of an electrode into any conductive material, in this case steel. Plasma burning uses a lot of electricity: short out the 100-amp service of the average house and create an arc, and the resulting power is about one-fifth of the current the Heath's single torch uses to cut one-inch steel. Obviously, where the arc occurs, heat is generated. By constricting the arc through a narrow orifice at the torch head and adding high-pressure gases, the arc becomes a high-temperature, high-energy plasma stream that hits the surface of the steel at temperatures up to 30,000 degrees Fahrenheit, consuming the steel where the arc touches it. The gas, forced out of the cutting head at high velocity, carries molten metal away, down into the water, leaving a square, clean edge.

The CM-56 oxy-fuel table beside the Heath works in a very different way. MAPP (Methyl Acetylene Propadiene) gas flows through the torch head, preheating the surface of the steel until it is about 2,200 degrees. The steel begins to spark a little, and when pure oxygen is fed through the nozzle, it begins to oxidize—to burn—very rapidly, giving off such high heat that adjacent areas reach kindling point and also spontaneously oxidize, creating a self-propagating burning path. Oxy-fuel burning is effective only for steel because steel is 98 percent iron, which oxidizes very easily in the presence of high heat and pure oxygen. The CM-56 can cut much thicker steel than the plasma burners, however, and it burns all of the plates thicker than an inch at BIW—one-and-a-half-inch, two-inch, and even three-and-a-half-inch used in parts of the rudder and for foundations underneath the very heaviest pieces of machinery.

For a flat piece of hull plate, the burning table may be the end of the line. It will be stored at Hardings until the Shell Shop at the yard is ready to weld it up, together with its structural members, before sending it into AB to become part of a unit.

After burning to dimension, all other pieces take one of two divergent paths depending on whether they are to be incorporated into an assembly large or small or remain essentially two-dimensional as a plate. Examples of plates are all of the flat expanses of deck, deckhouse, and bulkhead (what walls are called on ships), and the various flat and curving pieces of hull, called shell plate. Assemblies can be very small—any of the hundreds of different racks that are destined to hold some piece of equipment—or very large.

Most plates go to one of two places—to the Bending Floor or to the Panel Line. Welders reign supreme on the Panel Line, joining smaller plates into larger, usually rectangular panels, welding T-beam stiffeners in parallel rows that strengthen flat expanses of bulkhead, deck, and shell plate the way two-by-four studs support a Sheetrock wall.

The Hardings facility has presses—500-ton, 300-ton, 75-ton, 35-ton; it has two shears and an Ursviken framebender and a Metal Muncher punch/borer/shaper. The Bending Floor is dominated on one side by the hulking 500-ton press, on the other by an oil-fired furnace used to heat thicker steel until it is cherry-red, the color indicating a temperature appropriate for pressing. Next to the burning tables are the buffalo rolls—three steel cylinders, eight feet in diameter, mounted horizontally and running a good thirty feet across the floor. Thin plate steel is fed between them, then they are forced together to give the plate its rough curve before the last precise adjustments are made on one of the larger presses.

A whole bay—what the mechanics call "Galvo" even though nothing has been galvanized there for thirty years and the galvo pits were long ago filled with concrete—is given over to rows of individual workbenches. Each is manned by a partially disabled mechanic who works alone, using an ergonomically designed hydraulic lift table and specially adapted tools to fit up and weld, for example, a lifeboat rack out of thin steel or any variety of other small assembly.

Though it may not appear so, this is a place of great innovation at BIW, the fruition of the idea that getting back to work in any capacity as quickly as possible after an injury or accident is better than moldering away at home collecting workman's comp. In practice, a few of the men here are medically certified to work only eight minutes per hour, others have regimens that preclude any lifting, pushing, or pulling. Most of the men simply have bad backs, knees, or shoulders, and still function at a high level. They come to work every day, are not as prone to feel bitter, let down, or somehow diminished, and continue to be part of the crew as much as they are able. Some of them will finish out their careers here, some will find new positions more suited to their abilities elsewhere in the company, some eventually will return to their old jobs.

One of the most amazing—and highly respected—mechanics at Hardings is Cal Sutter, a gifted welder. Everyone knows Cal. Everyone's seen Cal, usually on hands and knees, a mirror stuck onto the end of a stick thrust into the farthest corner of an assembly, penlight in mouth. Sometimes the welder who did the work stands by, knowing that, if his hand drifted, pulling the electrode with its shielding gas too far from the work and the outside air got in, pockmarking the weld's face with minute wormholes and pinholes, Cal will find it. If he missed an interior corner, laid down a pass too thick or too thin in spots, Cal will gently point it out, suggest a few ways to fix it and avoid it the next time.

The welder may be chagrined at making a mistake, but it won't be the first time he's taken correction from Cal. In addition to deckplate welding and welding inspection, Cal is also the Hardings welding instructor. If the mechanic is new enough, chances are he's felt Cal's assured touch at his elbow before, during welding school on the third floor far above the deckplate, heard the supportive but plain-speaking voice in his ear.

Cal knows how hard it is to see from behind that mask when the darkened glass eyepiece that keeps the UV rays from frying your eyeballs flips down and it is *black*, when your body is contorted, twisted into an awkward position that you must hold endlessly as you swelter beneath the mask with its attached apron and the leathers that shield your chest, shoulders, and forearms from burning sparks. Cal knows the stench in your nostrils when the sucker hose pulling away the noxious fumes of burning metal gets knocked out of position, the occasional flash of claustrophobia inside a confined space lit only by the flare of molten metal at a couple thousand degrees above your head or at your feet.

Cal knows it because Cal does it. Cal is respected because he knows most everything about the "how" of the four kinds of very complicated welding done at the yard, and more than most need to know about the "why." And while in another company, maybe even another part of the same company, Cal would have been urged to find another kind of job, one perhaps not so physically taxing and demanding so much precise coordination, at Hardings Cal Sutter is not just another mechanic. He's indispensable, one of the few welders with the knowledge, ability, and willingness to move freely from teaching to doing to overseeing the work of others—always a sensitive area when one union member must pass judgment on the work of another. This is a guy whose identity has so little to do with his disability that, until you actually see him open a thermos, you don't realize he has only one hand.

The face of the future has shown itself here and there at Hardings. Entire bays have been shifted and reorganized to, for example, centralize the kitting area to speed the work stream throughout the sprawling plant. Out back, the railroad tracks have been replaced and certified so that the Maine Central Railroad can push railcars of steel directly into the yard while a new cab has been installed in the huge electromagnetic lifting crane out back, itself overhauled, to move raw plate from storage to the blast area even in high winds. Soon there will be a new highly efficient, accurate, and fast plasma burner, and the 500-ton press is up for a refit.

Hang a left through the heavy canvas drapes at the back of Galvo, and you walk into what is the heavy-industrial equivalent of an operating room. What used to be the malodorous and unhealthy Paint Shop has been transmogrified into the Powder Epoxy Line, a place of which Fred Hannah, the supervisor of the P-10s, or preservation technicians, is and should be immensely proud.

This is a square space perhaps one hundred feet on a side, one entire wall dominated by a two-story gray furnace off which the heat rolls in almost perceptible waves. On the left side Mike Atwater takes pieces from a wheeled table and hangs them from the hooks dangling from the circulating overhead conveyor system, which looks and functions exactly like a dry cleaner's circulating clothes rack except that its rate of speed is glacial, about four feet per minute. Each piece has been blasted with steel shot in one of the two blast cabinets across the room. The surface shines brightly and is slightly rough to the touch.

The conveyor snakes up and into the second-story oven, where each piece is preheated, depending on the thickness of the metal and the required *millage*—the thickness of the epoxy to be applied—to between 335 degrees and 525 degrees Fahrenheit. At the other end of the oven, behind twenty-foot-tall clear plastic walls, the pieces emerge into a cloud of fine dry particles of epoxy shot through spray guns angled to reach every surface, then cycle back up through the cure ovens. Back where they started, the pieces are sorted, stickered with a bar code, and received back into inventory, ready for further kitting, storage, or shipping to the yard.

The whole process is radically different from—and technologically light-years ahead of—the old days. When you preheat, P-10 Glenn Haggett explains, the heat melts the epoxy powder onto the piece. Also, inside the spray guns, you got an electrode. When your paint powder comes through, this charges the powder. We have a ground that we have to hook to the material to give it the opposite charge. It works like an electromagnet—the powder actually clings to the steel. So the paint melts on and you also have the electromagnetic effect.

In one corner a mechanic is wrapping about six inches of each leg of a large square steel table—a foundation for some piece of electronics or heavy equipment—with fiberglass tape before preheating and spraying. This is to save the lungs of the welder who, working on the ship, does not want to have to burn off acres of epoxy, releasing clouds of hydrogen cyanide gas, in order to have a clean surface to weld the foundation's legs to deck or bulkhead.

Atwater explains that this piece is just too heavy for the line, which can take about 100 pounds per hook, and so will be preheated and cured in one of the shipping-container sized "manual" ovens on the ground floor underneath the cure and preheat ovens after spraying in one of the vented booths. Beside each oven door is what looks remarkably like a large white kitchen timer, which is indeed the case. The guys used watches, Atwater explains, but then they'd get to talking, get interrupted, and forget the time. Gray epoxy overcooked turns green . . . which is not the color of ships in the U.S. Navy.

Before the Powder Epoxy Line was finished early in 1997, things were very different. He points to the foundation being masked. Used to be, Atwater says, once they'd got it all masked, they'd set it up on a table, they'd spray 'em first with green epoxy. That would have to dry for twenty-four hours. The next day they'd put the gray on, then that'd have to dry for twenty-four hours. So before, you're lookin' at a day of blasting and green paint, a day of painting it gray, and the third day they would receive them. Now we can do the same project—blast it, preheat it for forty-five minutes, takes about ten minutes to spray, cook it for another half hour, unmask it, and it's ready to ship. The entire procedure takes an hour and a half.

Hardings does still use the old way for the larger finished foundations. There is usually one mechanic in full-face respirator and white cotton full-body suit duct-taped closed at ankles and wrists to protect his skin from the noxious paint. He works in a special negatively pressurized, three-sided booth that sucks in outside air to keep paint fumes and particles inside.

To one side there is also a flame-spraying station. In this process, which is used sparingly to coat critical pieces like lifeboat racks that are exposed to the weather and must not corrode, a torch burns MAPP gas and oxygen, vaporizing aluminum wire fed through the nozzle. The high-temperature liquid aluminum spreads in a very thin, even layer over the piece, which will resist corrosion for up to twenty years.

■ ■ ■

Preservation Technician Mac MacKeown flame-sprays a lifeboat rack. The aluminum wire he grips in his right hand feeds through the gun, is vaporized by the torch head, and spreads itself finely over the rack like liquid aluminum.

B-Bay, at the opposite end of the building, where you walk in from the front gate, sometimes appears as a medieval vision of hell brought forward several centuries. A rash of small blue-green fires has broken out on the floor, sparks pinwheel out from menacing steel shapes, men in grime-streaked hoods wander about with strange instruments in their hands. All of this in a cavernous, dimly lit space where the air reeks of what could be brimstone and where the ears are assaulted by the peculiar deafening crackle created when a high-temperature electrical arc melts a stick electrode or wire into a seam to create a weld. The men behind the masks are working on large foundations, the structural steel underpinnings of truly heavy pieces of equipment like the 10-ton reduction gear, a gearbox about the size of a midsize car, which translates the high revs of a jet turbine into the various slower speeds of the prop shaft.

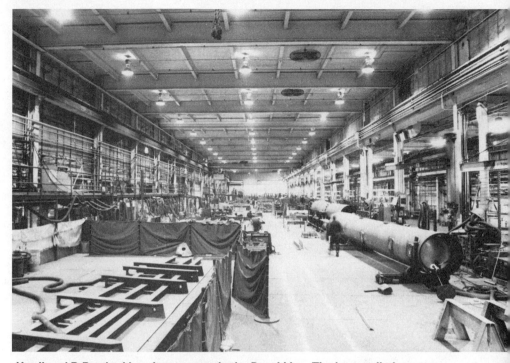

Hardings' B-Bay looking down towards the Panel Line. The large cylinders at right are lengths of propeller shaft tubing, and the lamp and sucker hose at the near end are for a welder working inside it.

Here the Vertical Launch System foundations come together as well. The most massive piece, weight for volume, that BIW creates, the VLS foundation is a box framework whose huge beams are built up from welded lengths of one-and-a-half-inch hard steel plate that must withstand the repeated assault of heat and stress as missiles are ignited within its narrow confines.

In brief, this is the work of Hardings, to incorporate smaller pieces into larger ones, to prepare stretches of structural steel for welding up somewhere else, to put the correct curve into shell plates for the hull, to weld up the large foundations for the heaviest machinery, to complete the smallest of foundations down to the final blast, primer, and final coat of smooth, flat, haze-gray epoxy paint. Every day, tractor trailers whose beds are piled high with these products stream down to the yard proper to be fed into the maw of AB.

Hardings produces only a few "finished" products, among them the beautiful matte-gray doors of watertight compartments, all of whose structural and moving parts are made there except the cast bronze handle. Hardings also fabricates the massive rudders, builds naval ladders, and has a corps of master welders who work with the so-called exotic metals—titanium, monel, cupronickel, for example—which are used for the twisting runs of turbine uptake and exhaust, shiny silver pipes as tall as a man that run from the ship's stacks down through the decks to the jet turbines that drive the ship and power its electronics.

Wherever the eye falls at Hardings, it is not immediately evident that a ship is being built. Even the mechanics sometimes don't know exactly what the piece they're working on is ultimately destined to become part of, to do. This sense of having taken a step back from the yard is reinforced by Hardings' real distance from the riverside ways and assembly buildings. Three miles is enough insulation so that Hardings has its own feel, its own way of doing things.

Don Lamson is the plant manager of Hardings. He's not particularly tall, and his dark hair is going gray. He dresses modestly in work shirt and dark cotton slacks. Without the white hat of management, he comes across as just another mechanic with a mustache. Watch him move through the plant, however, and his enormous authority is immediately evident, the respect he is shown as he stops to say hello, nod a greeting. Get him talking about Hardings and it's also clear that he is intimately familiar with his workforce and its strengths and problems even though he took over its direction less than a year ago.

Hardings is different, he says. It's like a big family here, and even though things are changing—they've been moving people around a lot in the last few years—we'd like to keep it that way. We do very little outsourcing, less than 5 percent of the ship, and we do 38 percent of all the welding that goes into every ship here at Hardings. The average age of our mechanics is forty-three, and most of them have about fifteen years of service in.

Sixty-five to seventy percent of my workforce has some sort of physical limitation, he says matter-of-factly, mostly cumulative trauma to knees, backs, hands, and arms. We'll never use pneumatic tools like we used to, all that edge prep and cleaning welds with chippers. He shakes his head, thinking of all the damage done to his men with those old tools, the old ways of doing things.

He begins to talk of the competition and of the challenge of the moment at Hardings, how to find the physical space and man-hours to increase Hardings' capacity by 30 percent in the year 2000 when BIW begins fabrication of its newest ship, the LPD-17 Marine amphibious assault ship. He explains that one of the measures of the efficiency of a shipyard is steel throughput, measured in tons per year. Kvaerner-Masa Shipyard in Finland and Mitsui in Japan, two of the world's best producers, move 50,000 to 60,000 tons of steel through their yards in a year building tankers and bulk carrier and container ships. BIW turns less than 15,000 tons of steel into approximately two DDG-51 destroyers each year.

With the additional work on the LPD-17 project, however, the yard must move the equivalent of 2.8 destroyers through the yard in the same time and physical space, and without a proportional rise in staffing.

Part of Don's battle will be won with the purchase of new machines, a few of which are already in place. Some economies have already been gained in reorganizing the storage and delivery of materials and by implementing a new database inventory control system, MACPAC, which uses bar codes and laser guns like at the supermarket checkout. The war will be lost, however, if BIW cannot change both the way the mechanics do their jobs and how they think about their work.

One man who is closer to his work than most is Peter Marshall, Hardings' blacksmith. Blacksmith? The first people making ships at BIW one hundred years ago were blacksmiths, he says proudly, and blacksmiths invented welding. Firewelding in a coal-fired forge where you get the steel up to near melting and by hammering it together you make a weld. Riveting is a pure blacksmithing technique, too, he continues, referring to pre–World War II days when ships' hull plates were joined by rows of rivets.

At six feet two inches, his gray/brown hair cut short under a powder-blue hardhat, well-muscled shoulders and chest swelling over a narrow waist and long legs, he even looks like a blacksmith. Completing the picture is a truly fine mustache, sprouting out—flying out, really—from under a generous nose and clear blue eyes.

Peter's world is a twenty-foot by sixty-foot railed-off section out by the burning tables, his space dominated at one end by the 100-decibel, red-hot blast of a small forge, its mouth spilling heat and flame. Butted up against the forge is what he calls his hammer and anvil—a 35-ton hydraulic press. This is a side press, meaning that the ram moves horizontally across the table from the right. On the left a wide heavy steel bar forms a stop. Using a variety of dies and pins around which steel can be bent, Peter "pushes" hot steel into different shapes.

He makes parts for the ship—U-shaped clips to attach ladders

Peter Marshall, Hardings' blacksmith, puts a chamfer on the end of a timber dog over the horn of his anvil. His red-hot forge glows in the back, left, mounted on wheeled steel table. The brick on top will cook his lunch.

to bulkhead and deck, bottle racks, resilient pipe hangers, and intricately bent dog cams only a few inches long that go into the locking mechanism of compartment doors.

He spends nearly as much time making things destined never to leave the yard, like the foot-long staples called "timber dogs." The two ends of each dog he delicately chamfers to a sharp pyramidal point so that foot-square keel blocks can be quickly and reliably joined across their end-grain by a single blow of the ship-fitter's sledge to support a section of ship being lowered by crane onto the ways.

Beam clamps capable of bearing loads from up to three tons are in constant demand. Because lengths of T-bar are often welded to the backs of completed panels of hull, decking, or bulkhead as stiffeners, Peter makes these two-piece, intricately curved clamps to fit around the T so that a rigger can attach a chain to move the piece around without the need for welding on—and having to gouge off—a lifting eye.

In some ways Peter is an anachronism, as are the hotworking techniques he uses every day, some of which have their roots in metalworking from the smithies of medieval Europe. I'm the last holdout of hope, he says wryly, that there is still a profound respect for the fact that things are done equal or better hot than cold. Everybody nowadays says, we'll get a bigger machine, more power, just squeeze it harder. . . . They've brought some new faces in to run some of these new machines. The old-timers knew what they *didn't* want to do; they respected the job I could do on a piece so they knew not even to try it in the first place. One of the big presses used to be a lot nearer to me and the operator knew over time what pieces not to do because they were too small—too intricate—where he had the potential to break a pin or just do a bad job on it because he couldn't get some of the corners. Some pieces I can shape better because I work hot.

Ladder clips and timber dogs are two things that, as hard as it has tried, BIW has not succeeded in making on a machine. Peter works them both hot, and with a lot of handwork; watching him is like stepping into a museum exhibit, until you realize the end product is not a Williamsburg rolled gun barrel, but a deadly high-tech warship.

Marshall straightens one half of a beam clamp on the table of his side press. Just shaped by the tall, straight die visible to the left, and glowing hot from the fire, the steel is still workable with a sledge.

A ladder clip is nothing more than a thick U shape. The ladder rail fits between the stems of the U, each of which has a hole for a connecting bolt to pass through. The curved bottom of the U is joined to the deck (or bulkhead) with a circular weld. The clip starts out as a flat rectangular blank of quarter-inch stainless steel two and one-quarter inches wide by three and seven-eighths inches long. The absolute first thing Peter does is to measure the length with a tape before using a small sledge and a punch to mark the centerpoints of the holes he then drills in both ends.

He carries a blank over to the forge, waves of heat rolling out of its mouth and off the top, dry in the nostrils and on your face. A mechanic walking by stops, removing his gloves to warm his hands at the forge. There is a single fire brick sitting squarely on top of the forge. He cooks his lunch there, pasta with sauce, leftovers, cheese sandwiches—wrap it up in tinfoil, throw it on the brick atop the furnace for a half an hour.

One of the hangar doors a hundred feet down from his space rattles open, and gusts of icy wind blow in. Peter gestures emphatically to back off from the forge's open end, out of which orange tongues of flame leap erratically, fed by the wind. That's my greatest fear, he comments, nodding towards the fire, being in the wrong place at the wrong time and getting burned pretty badly by the fire coming out at me. I can get away from it. He shrugs and turns back to the work.

He takes a pair of homemade two-foot tongs and picks up the steel blank, settling it in among the clinkers. As the paint flames up he turns his back on the heat, wiping the sweat from his face with a forearm. His eyes survey the orderly work space: the anvil with its protruding horn in the center of the floor, the racks of shelving behind the forge, hundreds of male and female die pieces stacked in piles, the clean desktop against the wall over which hang BIW calendars, framed photos of the destroyers and frigates he's worked on, dirt-blackened tables of squares and decimal equivalents long since committed to memory. Even his tray of finished work is neat and precise, beam clamps nested on end in regular ranks, ladder clips set inside each other making elegant S patterns like the border of a Caucasian rug.

He turns back to the fire, flipping the blank adroitly. Its edges have gone from black to gray to orange to a brilliant red, the soaking heat of the forge taking the steel from room temperature to 1,400 degrees in a handful of seconds. Black splotches appear on its surface, impurities called "mill scale" oxidized by the heat.

Peter's hands are protected by thick leather gloves, his arms from wrist to elbow sheathed in yellow knitted Kevlar gaiters. I used to wear welder's leathers with sleeves down to here—he touches his forearm lightly. I was macho enough to believe that all the burns I was getting on my forearms from the scale were part of the job. In my old age I discovered these—Kevlar. Ninety percent of the time I never get burned from the pieces themselves. It's moving or touching the dies.

He takes the red-hot blank from the forge, slots it horizontally into the waiting die, and pulls the handle to release the ram. The side ram forces the ends to bend around a pin, and a still-smoking ladder clip emerges, a perfect U with a nine-sixteenths-inch gap between the stems.

He moves with great deliberation and economy, short steps, precise motions of hand and arm as he takes up a small sledge and slips a one-fourth-inch blank—like the ladder end that will have to fit there—into the neck of the U. The cords of his neck strain as the sledge rises, falls, the flat sheets of back muscle moving under his T-shirt, thick forearms swelling as he hits the steel at the end of his tongs once, twice, on the flat of each stem so that the fit around the blank is snug. Finally he sets the clip on edge to see that it sits flat, that the bolt holes in either end will line up when some shipfitter out on the ways has to put the thing together. Thirty-five seconds to make a simple, but critical, piece, one with a bend so tight that it must be done hot because cold steel forced to curve so acutely would crack.

I have the dubious reputation of being wicked fussy, he says, smiling. The work I do is hot work—the small, the finicky, the finesse-type of stuff. There are some things, once you're set up for them, it's fairly brainless. When I make a hundred pipe hangers, I'm thinking about everything but Bath Iron Works. I build additions, buy brand-new trucks, make love to my wife fourteen times . . . Oh yeah, I screw up—daily, hourly, by the minute, he says, laughing. Obviously, with hot metal, you can mash the poop out of it if you're not careful, squash it from the right thickness. Making hammer marks on material is something I try not to do. There's times when I look at a piece that I've had to bang the bejeepers out of. Someone's gonna look at that and say, that guy doesn't know what he's doing, there's four hundred hammer marks on it! Think I'm an idiot, banging the crap out of the thing.

What Peter will never tell you, because he is too modest, is that he is one of the more respected mechanics in the whole of Hardings, the kind of guy whom visiting admirals invariably come to watch on a walk-through, who's never sick and never late, who gives a day's work for a day's pay every day. Though the product of private schools in Westfield, New Jersey, and Newport, Rhode Island, his dad a corporate lawyer in New York City, Peter looks like his coworkers, talks like them, if without any profanity and only the merest hint of a Maine accent.

With nineteen years seniority, Peter has worked almost half of his life at BIW. He makes about $16 an hour, time and a half over forty hours, and double time on Sundays. Overtime is two extra hours before the start of regular first shift at 6:30 A.M. three weekdays, and six hours from 6 A.M. to noon on weekends. Peter and his wife are Catholic and churchgoing, and Saturday afternoon and Sunday are special times to be spent with their four children. If he worked fifty-two hours every week—and overtime is not a reliable thing—his pay would top out under $42,000 a year.

They live in Topsham, eight miles down Route 1 from BIW, in a middle-class development full of two-story, four-bedroom houses just like their own. In the driveway sits a brand-new Ford F-150 pickup with the third door and extended cab so all the kids can fit. Inside the house all is neat and comfortable, if not luxurious. Peter has built an addition off the dining room, an impressive space of pegged post-and-beam, the wood glowing with a warm finish, angled buttresses finished off with chamfered edges and carefully routed detail.

Late on a Friday night, with a beer in hand and his teenage daughter and three young sons in bed, he settles back into the sofa, sighing contentedly. His wife, Valerie, is working the late shift tonight at L.L. Bean, something Peter is used to. Today, he says, I finished at three o'clock. Clock out. Mad dash. Always a big joke at Hardings: if you've never believed the dead can come alive, just watch this place at the end of a shift! So I met my wife down at the mall—she was going to work—and picked up the four-year-old, waited for the eight-year-old to get off the bus, went and got the six-year-old at day care, came home, got them settled in. I'm a twenty-first-century dad, he says without a trace of sarcasm. I cook and clean as much or more than my wife does. I hang a mean load of laundry.

Without conscious intent, Peter has kept his work well away from his home life. I can't actually say I have friends from Bath Iron Works, he says, a question in his voice. It's not that I wouldn't, but . . . when I come home, the focus is here.

I'm an implant, a highlander, a summer complaint. I was born in Poughkeepsie, New York, and brought up in Westfield, New Jersey. I'm so far from that right now. This is my home. This is what I do. This is what I am—I'm Peter Marshall, a blacksmith at the Bath Iron Works. Part of me says, I've started out what I want to finish; you want to dedicate yourself to some place as long as that place dedicates itself to you. And I think there's a corporate climate change evolving over the more recent years where that's not necessarily the case. We don't know now from one year to the next who's gonna own us or what their ultimate goal in life is.

I originally took a job at BIW because, I said to myself, I really don't want to be poor for a long time. Once I started doing metal-working there, I said, I'll just bide my time. I got married, started having children, bought my house. Okay, I have plans for a small smithy here in my mind, in my fantasyland mind. If I win Megabucks, it's gonna happen. Yeah, I'd rather be making funky gates and railings and this kind of stuff, but the other half of that double-edged sword is, wait a minute! Why can't I dedicate myself wholly to this? I help make billion-dollar warships that are phenomenal pieces of technology built by hand in a very proud tradition that's been going on for a hundred years. I'm part of that. Yeah, I'm a highlander; yeah, I'm from away. But I've been an integral and very knowledgeable part of BIW for twenty years!

A gray Saturday afternoon in March, and the Off Track Betting parlor in the basement of Bath House of Pizza is hopping. It sits just next to the yard, an off-shift institution where the kitchen help has long set their watches—and loaded the ovens—by the blast of the yard's whistle. Today the last of the spring snow spits down, and pedestrians slip and slither along the sidewalks, with the occasional glance up at the sky, the dim memory of sunlight and warmth in their eyes.

Kingsley Barnes sits at a table downstairs, his eyes on one of the dozen TV sets hanging from the ceiling, each with the name of a racetrack stuck to its base with Velcro. Aqueduct, Belmont, Scarborough Downs, and the pacers and trotters of the more small-time local tracks stream across the screens to the accompaniment of the familiar fanfare. There are serious bettors sipping beer and smoking distractedly, whole families where Grandma leads the cheers from her wheelchair, a barmaid with a bad dye job and deep lines in her face setting 'em up for the after-overtime crowd down from the yard.

Kingsley has just finished his Saturday overtime six-to-noon "up to Hardings." He sits and smokes his Winstons, much-scarred big hands flat on the Formica tabletop, smoke curling up past the square chin, knobby nose, making the brown eyes squint. Deep wrinkles radiate from the corners of the eyes and mouth, the cigarettes and sun and hard labor having aged him, like so many others at BIW, prematurely.

A twenty-one-year vet, Kingsley is the unofficial crew leader on the 500-ton press. He's a character, Peter Marshall says. Probably as redneck as you can get, but he's got a heart of gold. He's a good guy, and I respect him.

I didn't go all the way through high school, Kingsley says simply. I went two years, then I was diggin' bloodworms. Everything had froze up that winter, couldn't dig bloodworms or sandworms. I decided, well, I'll go to BIW for the winter, and I stayed six years the first trip.

For Kingsley it was no choice at all. Digging worms, bent over with a short hoe in wet sand for furious stretches while the tide was out in any weather—it was no choice. He came up the hard way, though he has no complaints. We were raised on a farm. There was twelve of us. We'd go through about a deer a week, eight boys, two girls, and Mother and Father. You didn't kill a cow every time you wanted something to eat. So we shot a deer. That's the way it was back in those days. Course, it's changed now. You probably couldn't get away with that. We used to. Born and raised on that farm, three hundred acres.

That's what I done at BIW for six years, was swing hammers. In 1971 the pay was $2.65 an hour. I was twenty-two. We worked on heavy fire in the furnaces. I was a striker—he has two hammers, and a guy moves a flatter, and you hit those flatters to shape the steel to the molds.

The Bending Floor is Kingsley's domain, a six-thousand-square-foot space whose floor is made up of inlaid squares of inch-thick steel grates. Steel not only because furnaced plates at 1,850 degrees would soon destroy the concrete by vaporizing trapped moisture in it, but because each six-foot square can be leveled individually, creating one vast straightedge. If you're straightening a twenty-foot length of prop-shaft foundation, you can use the floor as a reference point because you know that floor is level to within a sixteenth or a thirty-second of an inch from one end of the foundation to the other.

And BIW uses a perforated grate rather than sheet steel for good reasons, too. One is that steel is frequently cooled and shaped with water, which must run off somewhere. Another is the extensive use of dogs to hold things to the floor. One end of a dog, which is nothing more than a length of thick pipe bent into an acute angle, slides into the hole of the grate while the other rests against the plate to be worked. A few hammer blows lock the dog into the floor and against the plate.

This is the hot work Peter Marshall and Kingsley remember, how it was in 1940 when Hardings first opened as BIW manned up to produce destroyers for the war effort. Except that there were almost no men there at all. Many mechanics working there today have a mother or grandmother who banged rivets or worked hot steel at Hardings. Some things hadn't changed much by the '70s, when gangs of eight men would load whole flat shell plates, some ten feet by forty feet, into the furnace.

They would come out cherry-red, Kingsley recalls. We would bend some plate by heating 'em to 1,800 degrees, dogging them upright on the bending floor, then attaching several-ton weights to the upper edge and pulling and beating them over to the right shape with hammers. We'd set it up on the marks with the cranes and we would shape it to those marks with hammers by hand, 12-pound, 18-pound, we had some were 21 pounds. The 21's we'd use on the big heavy stuff.

I been burnt. Still got one mark on my lower leg, burnt right through the cloth. . . . The protective suits then? It was manmade material, when you backed up to them big furnace doors, it would shrink. That floor, the Bending Floor? I've seen it so hot we'd have to cool it down with water hoses. Just from dragging that hot steel on it. Heat it up, it'd burn your feet when you walked on it. Burn right through your boots, blisters big as quarters from the heat.

Another fanfare blares, and Kingsley looks up at the race, this one a half dozen trotters at Scarborough Downs, a track about forty miles south of Bath. Along with the announcer's commentary, Kingsley gives his own. He knows the jockeys, the owners, the horses, talking of them with a warm familiarity that betrays the longtime love not of a bettor, but of an owner. He's had racehorses for years, since he was a kid.

I race 'em at Scarborough, I go to New Hampshire, Massachusetts. I just bought a new place up at Pittston. I put in new barns, new paddocks; they get pretty well cared for when they're home. The colt, he's only five, not a really good horse. He pulled a tendon. I'm gonna get a vet to check him, get him froze-fired. What they do, they go in there and freeze it. You bandage it up, clean it until it reheals. See if that tightens it up. If not, I'll put him into a saddle horse. I don't wanna see 'im go to France—they eat 'em over there. I wouldn't be able to sleep good if I thought they cut his head off.

Kingsley is big, not tall—maybe a bit under six feet, but with a good 220 pounds on his ample frame. He's got shoulders and arms like a linebacker from so many years of wrestling plates into place, yet he can hardly close his fingers enough to shake your hand. Hang around shipbuilders long enough, and you come to realize it is one of the most common occupational hazards, cumulative trauma to fingers, hands, wrists, elbows, knees, backs, and shoulders from years of manipulating heavy tools and lengths of steel with too little rest and too much repetition.

Watch him work the 500-ton vertical press, however, and it's clear that his hands have a language all their own. The press is variously known as the "old girl" or the "old bitch," depending on her behavior. Black with the accumulated grime of welding smoke, metal dust, and years of leaking hydraulic fluid, she stands about fifteen feet tall and juts out onto the floor about the same distance. Viewed from the side, she's like a huge letter *C*, with two pistons emerging from the upper overhanging arm, each applying 250 tons of force onto whatever material lies on the rectangular table set into the bottom of the press's mouth. Materials are moved onto the table by means of two fixed cranes with swing arms, one on either side of the press.

Kingsley doesn't actually operate the press. He leaves that, out of deference and pure practicality, to Charlie DuBarr, who, at age seventy-five, has earned his place "pulling handles" on one of

Kingsley Barnes talks to the press operator with his fingers, telling him to bring the press head, the large square steel structure top left, down onto the work to straighten it.

the few chairs in the entire shop. Kingsley's job is to mount the dies on the table and ram head, to choose the attachments for stretching and rolling, to direct as many as three other mechanics when it comes to shaping some of the larger shell plates.

A rudder bearing housing is a simple, if critical, piece. Not only does the rudder that the bearings must support weigh several tons, but the force of the water flowing against it when the ship is maneuvering at high speed creates extremely high pressures against it. The housing starts as a flat piece of DH-36 mild steel a foot and a half wide by two and a half feet long with lifting eyes welded to each corner. It is three and a half inches thick and weighs more than 500 pounds. Scrawled in white paint, along with the part numbers, hull number, and various other information, is a simple instruction. Kingsley has to roll it up along its length to an interior diameter of thirty and a half inches.

First he sets the dies. A heavy steel template is rolled onto the table over steel pipes and bolted into position. Its center is a hollow trough with sides whose curve is exactly what the finished piece will require. Onto the ram head are bolted steel blocks that will concentrate the pressure in an area only several inches wide. With the housing blank set over the trough, the ram head will force the steel down, making it mimic the shape of the trough.

Lately the press, bought used in 1958 and due shortly for a refit, hasn't been functioning that well. Last week it happened twice to me, Kingsley says, I bring her down, and you call for your pressure, right? Nothing there. You take it up and bring it back down again—it might come and it might not. Takes two or three times.

Jim Anderson, the lead man (foreman) on the Bending Floor, and several mechanics have wandered over. Pushing three-and-a-half-inch steel is just about her present limit, and they're curious. Kingsley looks at them and smiles. You guys'll have to take a heat on it! he tells them. Everyone cracks up at this, a joke comprehensible only to a Bending Floor mechanic who learned on his first day that about the thickest steel you can shape with a hand torch and water stream is half inch.

Always take your hits from the outside in, Kingsley says. He means that, even though the curve will be the same at all points

Kingsley Barnes (right) and Jim Anderson (center) telling the press operator to back off on a piece being welded under pressure. The suited figure in the back left is the welder.

along the length of this flat piece, you start at one end, then move to the other, gradually working your way to the center from either end. This is how it goes.

Once he's satisfied the work won't move and that everyone is out of the way, he signals Charlie. The ram head descends until it hits the steel right on the mark, about four inches in from the end of the piece. Kingsley reaches one arm out so Charlie can see it, can see the finger clenched to thumb, reading from the speed of the clench and its force that Kingsley wants it slow, slow, softly softly, or more and harder. As the steel block presses into the housing, it begins to bend with a crackle, mill scale flying off it in sheets, the press whining louder and louder. Standing back from it, you can see the overhang from which the ram descends actually begin to move up, away from the work as the head pushes its way into the steel.

After every hit Kingsley motions Charlie to back off. The steel springs back a little as the ram rises, and Kingsley lays a half-moon-shaped plywood template against the curve, checking to see that the steel has been pressed enough. The piece is raised with the help of a jib crane, repositioned so that Kingsley can take a hit on the opposite end. After two rounds the housing begins to take shape, a long, flat U that will, hit by hit, curl up into almost a half-circle.

Kingsley can turn large flat plates into amazingly curved finished pieces of the hull. He also rolls up the thick steel for the missile silos, which are done in halves then welded together. This is a very precise, very tedious job that takes a lot of gutwork and sweat as you maneuver a long narrow piece of steel up and down the table, taking hits and measuring after each one.

We had one we had started, Kingsley says, and we got about half done. We'd moved it back and forth about six times—that way you can eye it up and see if you're coming good, keeping it pretty close all the time. One of the younger guys working with me says, Gee, you're pretty fussy with these, aren't ya'. Yeah, we want to be because they gotta be within one-sixteenth of an inch. I don't want mine coming back. It's work that *I'm* doing, I want it to be good. When those missiles come out of the silos, you don't want anything warped . . .

Back in the OTB, with the noise around him rising and falling with the races, Kingsley smokes, one eye on the monitor. Everything's changed a lot over the last ten years, he remarks. I used to work with the old guy on the 500-ton press, five, six years on nights. We were younger and the guy on nights had all the experience, so we'd go work with him. That's how it was.

You know how that five-minute whistle blows? Well, I'll never forget it. I was in there and he said to me, when that whistle blows I want you on the job, ready to work, your gloves, everything you need you had with ya'. And, he said, we work until the whistle blows for break, then we stop. Then he says, I want you back here before that whistle blows for break's over. You don't leave this job until the whistle blows for lunch. That's the way this guy was—and you worked every minute you was there. They don't do that today. Today everything is . . . Kingsley, his face almost puzzled, can't find the words and just shakes his head back and forth.

A lot of guys grumble about BIW, you know what I tell 'em? Go to McDonald's. You're gonna work in a place, might as well work where the money is. Most of us in there today is overpaid for what we do for work. If we could get the service we need to do the work, like cranes, if we could get the lifts when we need them? We'd be worth the money we're getting. They got three cranes in there, and you have to wait to get a lift.

I think we need to build warships just in case. I don't think we wanna get caught; you don't know who's gonna be the big threat. I watch the History Channel a lot, I watch to see what other countries are doing, like Hussein and them guys. I knew when we went over there the first time, they were gonna lose in a few days, desert-fightin' the United States! I was sorta joking around with a brother of mine. How long do you think this will go on? He says, well, these guys are top-notch. I say, what are they gonna throw at us, camels? I think the Navy should have a lot bigger and faster ships.

Always been at Hardings, Kingsley says reflectively. Only time I go in there—he jerks a finger at the yard next door—is if they have a lot of cleanup, painting, touch-up on finished ships. You'll see jobs that you done, all finished, and it looks pretty good. We do most all the big shell plates. And when you come down to the yard, walk along the ship, you can pick 'em out, the ones we done. This is one I had a hard job on right there. We used to come down a lot to drive wedges. We used to get a chance to see the job we done. It's almost over, he says quietly, one rough mitt absently stroking the black and white bristles on his chin. It's gonna be something in the past . . .

His words trail off, a fifty-four-year-old man with ruined hands and a fine mind, a lover of horses and dogs, a mechanic who knows how to bend and shape every plate in a destroyer's hull among a thousand other soon-to-be-forgotten skills. The construction of the new land-level launch facility with its detachable drydock is a sign, and one that Kingsley is smart enough to read. His time is almost past, and his expertise will be gone, no longer needed, rendered as useless as the soon-to-be-abandoned launching ways.

5 FOUR TO MIDNIGHT IN VULCAN'S WORKSHOP

Every trade has an unsafe act. I've nearly been flattened by side shell, taken out by foundations, had staging fall out from under me forty foot up. Been nearly blown up a couple a times inside a ship. You always wonder, when's that big one comin' at ya'?

—Mike Matthews, Shipfitter, AB

The BIW shipyard proper extends nearly three-quarters of a mile from the Carlton Bridge south, the river's edge dominated by the soaring cranes, the nearly complete destroyers moored alongside, and the scaffold-covered hulls rising on the ways. At the back of the yard, running along Washington Street, two hangar-like buildings squat, the largest a green-painted monster known to all as AB—the Assembly Building.

Running a fifth of a mile parallel to the river, AB is an order of magnitude larger than the Hardings plant, its hundred-foot ceilings and 100-ton cranes signaling that, indeed, large things are constructed here. The three sets of doors giving out onto the riverside roadway at intervals further the impression, each opening eighty feet on a side, their scarred paint and dented metal edges revealing that there are times when the riggers need every vertical and horizontal inch to move out what is inside.

On first shift, AB is crowded with more than two hundred mechanics—mostly shipfitters, welders, pipefitters, and tinknockers—as well as their numerous supervisors, all hard at work at any of the fourteen different workstations that line the floor on either side of a narrow central aisle running the length of the building. By the time the whistle blows for second shift, 4:15 P.M. to 12:45 A.M., the yard has emptied considerably, and AB with it.

Ninety mechanics more or less are assigned to second shift, and twelve of them are known as "Charlie's crew," a crack cadre of shipfitters and welders to whom are entrusted the assembly and pre-outfit of one of the most difficult and technically challenging parts of the ship.

At the far northern end of the building are the supervisors' offices and a conference room, rather grand names for what is actually a three-story stack of double-wide trailers laid out on I-beams. Standing three stories up on an outside staircase, looking way south out over the assembly floor, the main difference between AB and Hardings is glaringly evident: here eight entire recognizable slices of the ship known as *units,* many complete down to their keels, rear up at intervals along the floor while others take shape around them.

On the west side a bow unit rises twenty-five feet in the air, its keel plates resting on the shaped *mock,* or template, on which it was built from the ground up. The whole of its width is exposed, up top two decks with empty hatch- and doorframes, shiny pipe and ductwork gleaming, gray-painted foundations large and small sprouting up from floors and out from bulkhead walls in the exposed end compartments. The two decks are stacked on top of *inner bottoms*—what BIW calls the bilge spaces inside the curved keel that on these ships are used as fuel and water tanks.

Farther down the floor, on the east, or river, side, the main deckhouse rises three decks high, dominated by the four stop-sign-shaped SPY radar arrays set diagonally into its upper corners.

This is the 4240 unit, a "superunit" so called because it is, at sixty feet wide by fifty feet long and three decks high, larger in every dimension and heavier than most other units. It is also inherently more complex. All of the baseline radar information that the Aegis system uses for detection, defense, and attack comes through the unit's four arrays, which provide full-time, 360-degree radar coverage around the ship. The four arrays must be level *and* square in relation to one another within a half inch, and construction of the interior spaces is further complicated by the sheer amount of equipment and odd configurations of space inside that that equipment calls for.

On this night at the beginning of October, Charlie Barnes is worrying about the 4240 unit, worrying about going over budget because a surveyor screwed up. The unit has three full levels (called the 422, 423, and 424), three decks, and the bottom deck must be absolutely flat because of the vertical SPY radar arrays that are set across each corner. The thick deckplate that the four SPY arrays sit on is an area where the Navy demands "critical flatness." A deviation of more than half an inch means going back in, ripping it up, and leveling it. Unfortunately, Charlie's just learned from his fitters that the deckplate has had a slight hump built into it.

Charlie is, generally, a cheerful, calming soul who seems a good choice for lead man, what BIW calls "frontline supervisors." He's usually smiling, Charlie, a slight smile barely glimpsed from under a salt-and-pepper mustache and above a beard coming in more white than not. His dark eyes shine quietly under a pair of bushy eyebrows that turn up at the corners, giving him a slightly racy look. His face is otherwise dominated by a strong Gallic nose. When he talks, which is seldom, he never yells, waiting until the listener is close enough to speak into his ear rather than bellowing, like some.

Though this unit is Charlie Barnes's major responsibility, he is not alone in his worries. Unusually for a company where union and management are more and more divided, Charlie has a crew that is tight, motivated, and most of all cares about the work. The late-night visitor's first impressions are that, while they may joke and horse around a lot, they work like demons.

I've been lucky, Charlie says, with the new guys I've got. Course, they've been trained by Mike and Eric. Get trained by the right guys, get the work ethic in 'em, hopefully they'll keep it. Unless they get with a bunch that likes to sit down and twiddle their thumbs.

I had a welder here not too long ago—he was just lazy, period. Nobody on the crew liked him; he'd just sit on his ass and do nothing, and that didn't set too well. I talked to him a number of times, went and got the union rep. You guys got to talk to this guy, got to get him goin' somehow here or I'm gonna have to write him up. Yah yah yah yah yah. Well, this went on for four, five months. I started keeping a log, getting my ducks in a row to do my thing.

Luckily, they needed welders on the ways, my choice who to send. I suppose you could say it was the coward's way out of it—I pushed him up there. We have the right to assign. That's how I got rid of that problem. It can be like talkin' to a twelve-year-old kid. I'd just take him aside, say, hey, I been up on this unit four times tonight, you haven't been up there any of the four times I've been there. *Oh, I was in the bathroom or I was gettin' water.* I know better than that, because when I went up on the unit, I hung around fifteen minutes, and you didn't come back. *Well, I got this problem, I got that problem.* Understand, Mr. X, everyone's got problems, but you can't let it affect your work. I'm gonna have to try to help you by gettin' you some counseling.

When I can, I'd rather solve it face-to-face, him and I, keep the union out of it. Lot easier that way. With a lot of guys in here, you can't do that. They push you right to the limit, and then you gotta do the discipline. You gotta have your shit together when you do that, take notes—arrived on unit such and such a time, stayed here 'til such and such a time, welder was not on his job, over and over again. When it come down to it, there you go, guys, tell me what he *did* do tonight. He welded ten feet of quarter-inch fillet weld—all night! [An average welder laying in this simplest of welds would have an output measured in hundreds, rather than tens, of feet.] Eight hours! What's that tell ya? I was a fitter, I got a pretty good idea what somebody can get done in a shift.

Mike Matthews, a veteran shipfitter, wanders past Charlie, on his way back to his work site. Hey, Management Scum, he calls, we gonna get more time in the budget on 423? Charlie smiles, replies, No, and that's Mr. Management Scum to you. Now get back to work!

Charlie gives as good as he gets in the constant ribbing that is so much part of the way these guys communicate, but Mike gets away with more than most. This is because Mike is one of the most highly skilled and hardest-working shipfitters in the whole AB. He was handpicked to be part of the original crews who sat down, all shifts together, with the engineers and lead men to figure out how they were going to assemble the very first SPY array unit and then to tackle the daunting job of actually putting it together. Charlie knows, too, that Mike is the kind of guy who can never do just one thing at a time; he has been known to hold a conversation *and* tack-weld simultaneously.

Mike is in his mid-forties, and twenty-four years of the hardest of hard labor have taken their toll. He began his BIW career with two of the worst trades in terms of respect and working conditions: first as a cleaner, then as a tank grinder, removing rust inside the claustrophobia-inducing inner bottoms after water-testing. He's been shipfitting for two decades now, and the arduous work has left its mark. His unkempt gray hair flies out from under a much-abused hardhat, over a narrow, kindly face lined from both cigarettes and toil. Restless blue eyes and a fierce nose animate the face, punctuating the gray/brown stubble that covers his chin and cheeks. His hands, emerging from the too-short sleeves of a blue Marquette University sweatshirt worn to shreds, are chapped, cracked, scarred, and stained. The seat of his dirt-streaked jeans has been patched until there is no original material left, and his boots, glistening dully in the artificial light, seem to have absorbed all the grease they possibly can.

Used to be, Mike says, I never had enough vacation. For a long time it's been, hey, I'm not gonna take a day off—might miss somethin'! Cheap entertainment. One nickname I'm kinda glad died down is "Supahfittah" [Superfitter], which I got when some day-shift honcho was looking for one person from every trade on night shift who was outstanding in his field. So I got my picture taken with an admiral. Mike shrugs, a sour look crossing his face at the memory. Can't eat it, he finishes. Doesn't put money on the table. All it did was open me up for harrassment.

He enjoys good relationships with almost everyone, helps to break in the younger guys on the crew, and keeps up the morale—and the pace—by his sheer spirited presence. He can, and does, step in as backup lead man at a moment's notice when Charlie is sick or on vacation, and often doesn't know from one day to the next if he'll be wearing Charlie's white hat or his own, much-preferred, brown one.

One of the first things I tell a new guy, Mike says, is, you had your orientation? Forget everything you heard, and we'll start from scratch. How do you like your coffee? And I give 'em the ground rules for the unofficial coffee break: don't hang out on the floor, take your cup back to the job while you figure out what you're going to do next and you'll never get in trouble. They show you the black-and-white rules, but they don't tell you the socially acceptable rules. I like to try to make a new guy comfortable right away, mold him into the crew. He's an outsider until we all decide to accept him. The sooner you make him comfortable, the less he'll get hurt, the more he'll listen to you, and the less likely he'll hurt someone else. Another thing I tell 'em: don't develop bad work habits like some people here. They pay you to work, so work!

At the same time part of Charlie's team is finishing up this three-deck array unit at the north end of AB, a handful of others at the south end, where the assembly process truly starts, has just begun on the 423 deck of another array unit for the next hull.

To visualize the construction process, imagine a finished ship cut into forty-foot-long cross-sections, then each cross-section divided again horizontally into subsections one deck—about nine and a half feet—high.

The ship is divided into about seventy of these subsections, which are called *assembly units,* and they tend to be large. Each represents a slice of the ship about fifty feet wide, forty-eight feet long, and usually one deck in height. (See Chapter 2, Figure 7.) As each unit moves north through the building, it gets more and more complete and is then mated to a deck above or below to become an *erection unit,* of which there are forty. When it goes out the door headed for the ways, the unit has gone from duckling to swan. It's still nearly fifty feet long, but can be as wide as the ship at its widest (about sixty-seven feet), and up to three decks in height.

The SPY array unit, officially the 4200 erection unit, starts life as four assembly units, each one deck high—the 4220, 4230, 4240, and 4250, though in everyday speech the mechanics call them "the four twenty-three," "the four twenty-four," and so on. Each level is assembled, welded up, and pre-outfitted as much as

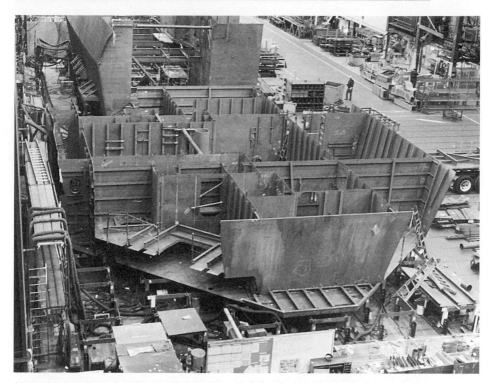

Here the octagonal 4230 deckhouse base is initially fitted up when inverted. The four SPY radar array panels will be set diagonally into the corners.

possible; then, still inside AB, the 4220, 4230, and 4240 are joined to each other. The topmost, 4250 assembly unit, a small steel rectangle about thirteen feet by thirty feet, though only seven and a half feet tall, would make the structure too high to fit out the door of AB. It is erected after the whole unit has moved to Five Skids, the open area next to the unused C-Ways where the tallest deckhouse units are finished beneath billowing nylon tents.

As it is, some of the units will barely fit out of the doors of AB. Weighing up to 200 tons, a few also approach the safe combined lifting capacity of the two 100-ton inside cranes that must hoist them—and sometimes the mocks on which they were built—into the air so the wheeled transporter can move them to Five Skids, or to Blast and Paint or Pre-Outfit 2, both buildings just across the roadway from AB.

Randy Fisher and Luke Arsenault, both shipfitters, are working on top of the 4230 assembly unit, a flat, octagonal expanse of heavy deckplate whose face is crisscrossed by a network of supporting deck longitudinals and heavier transverse beams. The deckplate is so thick—more than an inch in places—and so heavily reinforced because of its function: the 4230 is the base unit under the SPY arrays, and as such must not twist or vibrate in heavy seas, nor fail catastrophically under the immense pressures created by the explosion of an incoming shell, missile, or nearby nuclear blast. Should the SPY system be destroyed, the ship would be nearly blind to attack and greatly hampered in fighting back.

The deck sits up off the ground about four feet on a *mock,* a forest of steel stanchions adjustable in height about fifty feet on a side. The surveyors have been through and set the heights so the deck is level, and first-shift fitters have tack-welded it in various locations to the stanchions to hold it in true against the normal distortion that welding will produce in heated steel when the bulkheads are laid in.

Randy is a third-generation BIW mechanic. My grandfather, he'll tell you proudly, led an all-women crew up at Hardings during World War II. I got an aunt who works on a framebender up there, two uncles down in the yard, I've got relatives working here I've never even met. A David Crosby look-alike with a little more hair and a burlier frame, he's soft-spoken and shy, his features rough and kindly: a gourmet potato of a nose, red-burned skin, laugh lines around the bright eyes and mouth. He looks like he'll make someone a great granddaddy in about ten years.

He's not long back after surgery to repair vertebrae in his neck destroyed through years of cumulative trauma. In his case, it was in large part the welder's head snap seen everywhere around the yard, in which, with the work set up but both hands occupied, the welder lowers his hood over his face by jerking his head down.

Now Randy doesn't generally do more than tack-welding, the shipfitter's hammer and nail. A tack is an inch or so of rough weld, usually stick-welded, used to join two pieces of steel temporarily in the assembly process. In joining a long bulkhead to the deck, for example, the fitter will put in a tack at one end, then move along in intervals, checking that the deck is level, pulling down high spots and pushing up low spots, throwing in tacks as he goes. If he has made a mistake or runs into a problem requiring removal of the bulkhead, it's neither difficult nor time-consuming to gouge out the tacks and begin anew.

Randy used to be the guy who followed the fitter, the welder under the hood on hands and knees "bailing that wire"—covering the ugly tacks with a continuous even ribbon of pulse-arc weld. He's still allowed to advise, however, and you could have found him earlier in the evening at the side of Matt Larochelle, whose black hardhat has the new-car shine that identifies him as a rookie. The older man kneels down, chalk in hand, over a long horizontal weld built up of many passes joining a thick exterior bulkhead to the deck. Good job, he says, you're gettin' some-where. He marks up the weld, pointing out spatter that needs to be ground off, a low spot to be filled, an unacceptable divot where one pass ended and another began. His words are slow and easy, his manner matter-of-fact as he diagrams a fix with his chalk on the bulkhead above. Matt starts to explain something about first shift did this or that, and Randy shushes him, passing on some advice more homegrown than technical. Here, if you make a mis-take, tell us. We'll figure out how to fix it. Don't blame it on first shift or someone else. Everyone screws up, he says, shrugging.

Reborn as a shipfitter, tonight Randy has been doing prep work for laying in a bulkhead. First he makes his way slowly along the white inked lines laid out to show where the bulkheads will run, his hammer hitting the head of a punch with a rhythmic clang over and over, every few feet. The punch, made of hard steel, makes a clear impression in the deckplate that will remain should the ink be removed or covered or the underlying weldable primer burnt off. Next he grinds areas of deck down to bright metal where attachments once were welded on, so the bulkhead will fit cleanly to it, and strips rust and slag from the ends of the T-bar stiffeners so they will take a weld cleanly and well.

The last step is trimming the ends of stiffeners and beams that protrude into the path of the bulkhead, and grinding them as well. He picks up a plasma torch, adjusts its flame until a blue pearl glows at the tip, then begins to burn off a slice of T-bar, his hand rock-steady, his body relaxed, the slices of unwanted steel falling away neatly and regularly, leaving a smooth edge that will need just a touch of the grinder. It's not hard to see that he is still a very good welder.

The 423 unit they're working has come up to them, craned over the low wall that separates the Panel Line at the far end from the rest of AB. This is the flow of work in AB. Steel panels destined for shell, deckplate, or bulkheads are burned to size and made up by Hardings and delivered at the most southern end of AB. From then on in, the movement is north through the various stations of AB, until the unit goes out the doors just down from the ways, every unit's ultimate destination.

The Panel Line is the first stop, a place where smaller sections of flat plate are welded together into larger ones before having the various stiffeners and beams fit up and welded to their backs.

The second stop is one of three Main Structural Assembly stations, where the bulkheads—the interior walls of the ship—are "hung" and welded to the deck, and to each other where they meet. It is here, in MSA, that the visitor can see and instantly grasp the logic of what was originally a major BIW shipbuilding innovation: *downhand inverted unit assembly*—constructing as much as possible of the unit upside down.

Inverted assembly just isn't possible for all assembly units, and towards the northern end of the building are a series of mocks for right-side-up assembly. These tend to be units where, as the mechanics say, the keel has "a lot of shape" to it. Larger bow sections, or near the stern where the shafts emerge from the hull, are both examples of lots of shape. Unlike the flat mock of the 423, the mock for a bow unit will mimic the curve of the hull at that point, a deep V, and assembly will start with sections of shell plate—the hull—being laid in and pulled down onto the mock to the correct shape. The shell will be held in place with tack welds to the mock itself, and assembly will then proceed from the keel up.

Watching mechanics clambering around trying to fit and weld on the steeply sloping insides of a bow unit, you might think that there should be an easier way—until you consider that the mechanics would have to fit up and weld those same curving, unwieldy, and thick plates last, twenty or thirty feet up on top of the unit. The rigid support of a mock ensures that the hull will take its intended, designed shape, a factor crucial for the ship's speed, performance, and stability in the water.

The 423 is an inverted assembly unit, and the upward side—what sailors will walk on—is the side now facedown on the stanchions that make up the mock. The mechanics are in effect walking around on the ceiling, working in what is called the *downhand* position. Since so many of the major systems—electrical, ventilation, lighting, piping—run overhead, it's far easier and more efficient to install them with the unit upside down when the mechanic doesn't have to perform every task uphand—with arms upraised.

It's easier on the body, Mike Matthews says, to work that way; you get a better quality work. It's usually quite a bit cheaper, fewer fixes and repairs. For one thing, uphand puts you in a position your body resists—that drag, gravity, you feel it after a while. As far as doing your repair work, anything that involves climbing up and climbing down—it's time-consuming, too.

Though inverted assembly may be efficient, it's not easy and it brings its own problems. As Mike points out, we have a cover plan, which shows us everything we're gonna do on this unit, where we're gonna put all these bulkheads. And it's gonna be a nightmare to get around on [the unit] because every one of these bulkheads is gonna have a doorway in it, and, because it's upside down, the doorways are gonna be coupla feet off the deck, so we're gonna have to stage up every one to climb over it. We're gonna be running all the piping, vent down close to where the ceiling is gonna be, a long process.

Usually with bulkheads, he continues, we start with the back, then we do the sides, then the front. We'll be dangling plumb bobs off the unit and reading the angle of the dangle. It's gospel here that we got nothing square above the waterline; everything goes on 10-degree angles for radar deflection. So any time we got something above the 01 [main deck] level that seems plumb or square, we question it, Mike finishes.

There aren't a whole lot of different trades at work here, as evidenced by the fact that most mechanics wear either the brown hardhat of a shipfitter or the black hat of a welder. In fact, the

MSA crews operate like crash framing crews, with the other trades waiting, like subcontractors, until the basic structure is in place. On Charlie's crew, the fitters often work in pairs fitting and tack-welding the bulkheads, with the welders following right behind them, laying in continuous welds over the tacks.

Three mechanics maneuver a long, hanging bulkhead into place, lining up the notches in the bulkhead's bottom edge to fit over the longitudinal stiffeners of the deck. The taller T-beam parallel to the bulkhead at their feet is a supporting transverse beam necessary, together with the closely spaced stiffeners, to support the great weight of the arrays and lend the rigidity required for the radars to function accurately.

Randy grinds and trims until after supper, when the first bulkhead is ready to be hung. The crane's siren blasts over the Bulkhead Pre-Assembly area just across from the 423 unit, two orange-hatted riggers scramble off the steel, yelling to a welder on the next bulkhead over to move away as the chainfall tightens, pulling the bulkhead vertical. This bulkhead is a single panel about thirty feet long and ten feet high to which has been welded ranks of horizontal T-bar stiffeners like two-by-four studs. Panel and T-bar together weigh a couple of tons. As it tilts slowly on end, objects—slag, scraps of steel, bits of wire and welding rod, a forgotten hammer and punch—begin to rain down with a boom and a clatter where the welder was just working.

As the crane moves the bulkhead over to the 423 mock, one rigger walks along behind, guiding it with a rope trailing from one corner. Randy is up on the mock, and he and the two riggers manhandle the bulkhead until its notched bottom edge is hovering a few inches above the tops of the neat rows of stiffeners and beams it must fit over. The crane lowers it slowly, and it gets hung up on the lip of a beam. Randy bashes it with a maul, and the end falls into place with a bone-jarring crash.

Randy quickly puts in place a bulkhead brace, a heavy steel pipe with small *pad eyes*—four-inch squares of steel scrap with a hole cut out of the center, which the fitters refer to as "assholes"—on swivels at each end. He tack-welds one pad to the top of a deck beam several feet from the bulkhead so the brace meets the bulkhead at a 45-degree angle, then tacks the other pad eye to the bulkhead. He repeats the process with more braces farther along the bulkhead, then stands well clear as the rigger unhooks the bulkhead.

I've seen a bulkhead fall over on a guy once, Mike says. Bulkhead slipped. It was a lot bigger than he was, and he's trying to hold it up. I'm screaming at him to let it go and just get the hell out of there. And he's trying to hold it up! Finally he got out from under it and let it drop. Big noise. So what the hell was you going to do when you got it upright, I asked him. Did you think that thing was going to balance on edge? It was going to go over the other way! He wasn't thinking. You gotta learn to think all the time here.

With the bulkhead in place, Randy and Mike begin the tedious process of making it up to the deck, putting in an inch or so of tack weld every running foot starting from one end. From here on in, it's a game of push-pull with the steel. For pushing, they'll use a 10-ton jack to force one bulkhead up to another or, if a section of deck has a sag, a heavier 35-ton jack underneath to force it up. For pulling, they'll use either a light *cable lug-all*—a portable, hand-cranked, geared winch with a hook at both ends of the cable and a handle to reel it in—or a *chain lug-all*, whose heavy links can take much more tension.

Later that night Mike is crawling on his hands and knees under the mock, dragging a chain lug-all behind him. He's already got a stick-welding rig and two pad eyes under there. He is working with Merle "Whiteass" Witas, a chunky young guy who responds to Mike's constant jibing with a gap-toothed smile. Merle's up above, where the two have been attacking a high spot fifteen feet long where the deckplate is bulging up, creating gaps under the bulkhead edge and making it sit like it's a seesaw.

Underneath, Mike first determines where the bulkhead above runs and marks it with chalk, then tack-welds a pad eye along his chalk mark. He upwelds, and sparks shower down on his forearms, setting his sweatshirt briefly on fire. Smoke billows off the end of the welding stick, and with nowhere to go, rolls down off the ceiling in lazy spirals. Go on and lay in some weld, there, Merle, Mike calls up through a hole in the deckplate. Next, he tack-welds another pad eye to one of the flat steel bands set into the concrete floor for that purpose, strings a chain lug-all beneath them, and begins to crank. Merle's got to put enough weld on the bottom of that bulkhead—eight inches or so—so I don't just rip the tack welds off it, Mike explains, grunting as the chain tightens, pulling the deck down towards the floor.

Lot of these jobs, he says, granted, one guy can work it, but it's not cost-effective. You lay a lug-all in underneath to pull it down, then crank it, run up top—am I there yet? Jump down and crank it some more. With two people, one under, one up top, it saves so much time running. We have a hammer code, some of us—one bang, crank it more; two, good; three, back off.

Suddenly the pad eye rips off the strapping, flies up and hits the deck, then clatters to the floor. The close space underneath booms with the sudden release of tension on the plate. Jaysus! Mike yells, jumping back. His hands were only a few inches from the path of the pad eye. That was not funny, he says. I coulda knocked ashes into my coffee!

Every trade has an unsafe act, he observes, and this is ours. Heavy metal works hard, and a lot of this deck, it's inch-and-an-eighth plate. What I do, I never clean a weld when I put on a pad eye. And when I'm cranking, I'm looking at the slag on that weld. When it goes—begins to crack off—I go, too, he says with a grin, 'cuz I know it's gonna come off! Randy and the other fitters, who've heard Mike on this topic a time or two, tend to think he's just plain full of shit on this point, having seen their share of welds fail without any warning at all.

When one bulkhead is in and it's time to make up another one to it at right angles—forming a T or an L, for example—Mike is especially careful. What I need to be concerned about is, when I make up bulkheads vertical to vertical, a T shape, it will lock it. If the deckplate was high, she'll lock high.

That's what happened with the nearly finished array unit Charlie was worrying about. The deck had a high spot in it, onto which the bulkheads were welded as if it were level, "locking in" the crown—carrying the mistake forward into the next stage of assembly where correction becomes more and more difficult. So, Mike finishes, we're gonna know if it's level. Sometimes we'll even get the surveyors down here to do a spot check.

Two weeks later the 423 unit has miraculously risen. What was a flat expanse with a few verticals is now recognizable as the base of the main deckhouse with its characteristic octagonal exterior bulkhead walls erected. The mechanics have covered most bulkheads with graffiti ("See Harold Work—50 Cents. No Refunds!") above an opening that looks like a window. What is not graffiti is an inspector's terse chalked directions, which seem to tell a story of things gone wrong: "OFF! On Wrong Side of BHD!" "Weld Hole!" "Do Not Make Up BHD Until DK Goes Down!" "DNW—Do Not Weld" "Do Not Remove Jack!" and an arrow pointing to an empty mount where a jack used to be.

■ ■ ■

After a full month in MSA, with most of its bulkheads in place and its rough tacks covered by brightly gleaming, even ribbons of weld at most joins, the unit moves north again, this time to one of the High Velocity Outfit stations (HVO). Before it is moved, two things must happen. Overnight cleaners swarm in and clear every flat surface of debris, collect every steel scrap and mislaid tool, and sweep up and remove every bit of trash. They are meticulous, and their work is carefully inspected. Past experience shows that the main danger in moving units is not a lifting cable's parting or other equipment failure, but forgotten tools and scrap steel falling off in transit onto unsuspecting parties below. The stagebuilders come next, unhooking portable stairs, removing ladders, taking apart the scaffolding, generally making sure that when the unit rises off the floor, it won't take anything with it.

Finally the riggers move in, two of them, to direct the 100-ton crane. Every move is unique, and each is carried out according to the dictates of the *Book of Lifts,* which diagrams each lift, what length wire-rope slings, how heavy the D-ring shackles, where each shackle is attached to the unit. At roughly the four corners of the unit, welders have attached 100-ton lifting pads, which look like doughnuts the size of small tires with one flat edge welded to the deckplate. The D-rings the riggers use are as long as your arm, linking the wire-rope slings above to the crane's dangling hooks and below to the lifting pad. With the great stress of lifting, the pads often must be attached to the thickest, most well supported part of the unit, often the underside of the deckplate. Because the deck's sharp steel corner edge would fray and perhaps snap the individual thin woven strands that make up the sling, the edge has been capped with a length of *roll pipe,* a rounded lip made of heavy steel.

Once the rigging is complete, the rigger uses his lapel mike to communicate with the crane operator, and the slings tighten. The unit comes almost invisibly free of the mock, swaying and shimmying a little, rises at an even but glacial pace until it hovers twenty feet off the floor, then moves north at a walking pace to the High Velocity Outfit area, where a bed of pipe supports has been laid to hold it about a foot and a half off the floor. Unbelievably, a rigger walks under its edge (his trade's unsafe

The octagonal 423 array base descends onto a bed of short pipe supports in HVO while a rigger, far right to the rear, walks the perimeter to check for obstructions. Roll pipes are clearly visible both at the bottom edge of this unit where the sling goes under it and on the top edge of the unit just to the right.

act), then parades the perimeter as it descends, making sure no staging or other material protrudes into its path.

Thirty minutes later, with the slings detached and ladders up, Mike's nightmare has come to life as the unit is invaded by more trades—pipefitters and braziers—who will now compete with Charlie's crew for time and space to get their job done. Typically, two further trades with their own tools and parts, tinknockers fitting duct and even electricians running lights, may also complicate the picture.

At times as many as fifteen mechanics crawl over, under, and through the unit's interior maze-like welter of small rooms, half walls, overhangs, and upside-down doorways. If you didn't know what they were doing, you might think it a steel jungle gym created especially for the amusement of demented mechanics. The whole unit is overhung with the fat snakes of sucker hoses, red pressurized air lines, and black welding leads running from the pulse-arc machines to the suitcases containing the rolls of wire that are fed at a constant rate through the welder's electrode.

From above, the unit looks like a house with its roof removed, except more properly it is the floor that is gone and you're looking at the ceiling. The top edges of the bulkheads, not yet straightened, ripple and careen left and right; and almost in the unit's center, a round steel column rises in an open space across whose top runs a thick stabilizing beam. Pipes as small as a little finger and as large as a man's thigh, for cooling water, fuel, heated oil, oily waste, and sewage lines, run in bizarre patterns underfoot, jumping some foundations and going through bulkheads, snaking over stiffeners and beams before they end abruptly in space at the edge of the unit, their ends carefully capped and labeled.

Every sound is magnified inside the steel labyrinth, the screech of Merle's grinder leveling off welding scars, the rhythmic clang of Randy's hammer falling on a punch laying out a bulkhead, the background thrum of the huge ventilation fans overhead, the crackle and pop of Luke stick-welding the next bulkhead over, the hissing rush of a plasma torch cutting through steel, the characteristic electrical buzz of a welder's pulse-arc gun melting the endlessly uncoiling wire into a seam, the deep, complaining bongs as Eric tightens a bolt, dragging several tons of steel unhappily into alignment, the whole overlaid with the staccato shouts of mechanics struggling to be heard over the cacophony.

Even with the yellow foam earplugs everyone wears, a hammer blow on steel plate inside the unit—with nowhere for the sound to go and every bulkhead a reflector—is palpable on the eardrums, a hot wire of sound felt as pain. And this is not the only danger. Every step, and your foot meets a potential trip hazard, every unfinished steel edge reaches out to slice your skin. The safety inspectors tag sharp corners with yellow safety tape whose average life expectancy is a shift or two before a gust of wind or a wayward tool rips it off.

It doesn't take long to realize why everyone wears so many layers of clothes here, either. The cavernous building has no effective heat, and one of the huge doors is almost always open as units and bulkheads and forklifts bearing pallets of parts enter and exit, letting what warmth the work has generated dissipate. The steel itself is cold, yet gloves are necessarily thin so that the mechanic can manipulate his tools. Mechanics working anywhere near an open door are subjected to icy blasts of wind off the river. Rain and snow blow in, and it's not uncommon for any standing water left overnight to freeze up, especially in the depths of winter when the outside temperature can hover at 10 degrees to 15 degrees for days and even weeks at a time. We get our own weather in here, Mike says. Summer, sometimes the building's so thick with smoke you can't hardly see one end to the other; and winter, it's been known to snow inside the building.

While the mechanics layer up against the cold, they also do so in self-defense against cuts and burns. Since they work at such close quarters to each other in the small interior spaces, it's not uncommon for a welder up top to be unintentionally raining down hot sparks or slag on a fitter below. Mike says, coughing into his hand, it's been known to happen on the array unit, when fitters and welders find their space invaded by too many other trades, for one of them to do a little carbon arc gouging. The deafening machine-gun chatter of the pressurized air, combined with its great heat, intense light, and profuse sparks, encourages the pipefitters and tinknockers to get the hell out of there, and fast.

You can always recognize a welder by the pinhole burns that make his sleeves look like Swiss cheese. Even the hardiest work clothes, if not dotted with pinhole burns from welding, burning, or grinding, can be reduced to shreds anywhere inside the unit, where every unwelded edge is razor sharp, especially where it has been freshly burnt, leaving a ragged burr on the side away from the torch head.

Among the fitters and welders of Charlie's crew, who spend most of their time crawling around on the filthy steel for eight hours every night, Eric is the exception. In sharp contrast to the others, Eric always looks as if his clothes began the shift not only clean, but pressed. He's a sharp dresser—jeans that fit, bright sweatshirts with no holes, his face freshly shaven every day. In his left earlobe, a small diamond stud flashes above a jeweled hoop. He's a little under six feet, squarely built, and with the thick muscles of a bodybuilder bulging under his sweatshirt. He wears his brown hair short and cut neat, his mustache is clearly trimmed daily, and it's only when he takes his hat off that his bald crown is revealed.

If Eric is the tackle, Mike is the fitter's equivalent of a tight end. Eric is methodical, strong as a bull, not afraid to use brute force. Mike will often sit back, do three seemingly unrelated things that come together to accomplish with two beam clamps, a wedge, and a lug-all what Eric does with two welded-on pad eyes, a jack, and a sledge. Mike makes a game out of it, wanting first shift to come in the next day and wonder how the hell he got a complicated interior foundation up without using a single welded attachment. But Eric knows how to weld, too, while Mike, like most fitters, is considered by the welders to be a steel butcher, a terrible welder without art or skill.

This night Mike and Eric are working on either side of the same thick exterior bulkhead, Mike up top and Eric down below. Inside the unit, Mike wrestles with a blueprint that sketches out the pieces of a small box-frame foundation that has to be attached almost eight feet up the side of the bulkhead. The original pieces, cut by Hardings, proved when unkitted to be totally screwed up. So Mike's making a new foundation from scratch, rebuilding it piece by piece from scrap stock.

He is perched in the corner of a four-foot by ten-foot rectangular compartment whose every surface is cluttered. For tools he's got a four-inch plasma cutting rig with its leads, kneepads, C-clamps, end wrenches, electrical tape, leather gloves, a flashlight, a carpenter's folding rule, one small jack, one chain lug-all, one 30-ton jack, two wire-rope lug-alls, a tray of welding rods, leads, the welding electrode for stick welding, a face mask for protection against chips and sparks when grinding, a hood with dark glass eyepiece to shield the eyes against UV burn when tack-welding, two toolboxes, a pressurized air lead, and a couple of beam clamps.

He chalks lines on the rough stock, then burns them out with a plasma cutter, each piece falling to the deck with a clang. After grinding their edges quickly and deftly, he gathers the pieces and clambers up the side like a monkey, using the bulkhead's horizontal stiffeners like steps. He's working high up, one foot resting on a light hanger sticking up three feet from the deck, the other on the protruding lip of a stiffener.

After using C-clamps to fit up the pieces and rechecking with level and measure, he shoves a stick between the jaws of his welding electrode and dons his leather gloves. Holding his hood in front of his eyes with one hand, he begins to weld, legs spread precariously, tap tap tapping his stick into a corner until it arcs and burns blue. He hops down to take up the plasma cutter, trimming a scrap from one protruding leg, then grinds it clean before tapping another piece into place with a ballpeen hammer.

Down below, Eric is manhandling three-foot by five-foot steel panels onto a protruding frame with much crashing and banging, knocking Mike's piece out of true. Mike pokes his head up over the side of the bulkhead. Hey! Whatchadoin! he calls, throwing a fistful of foam earplugs at Eric playfully. I'm tryin' to work up here. He ducks as a piece of trash comes sailing back. Then he calmly picks up the fallen piece, taps it into place, and grabs up his electrode again.

Eric's panels are warped from the heat of being cut, and he yells for Merle to help. Come over to bail your ass out again, Merle chides him. Even the two of them can't get the panel flat and still weld it until Charlie sees what's going on, hops up on one end to hold it down with his weight. Eric lays in a tack, then Merle whacks the weld twice with a heavy sledge. That's called a "hot tack," Eric explains. When the weld is still hot, you can compress it with the sledge, taking the gap out and getting a tighter fit between two pieces.

Midnight on a Friday night and the pace slows perceptibly, then stops altogether. Tools are stowed and leads coiled, hands washed and bladders emptied. Most of the mechanics seem to have mysteriously vanished, and the cavernous building, nearly quiet and almost empty, takes on an otherwordly feel—Vulcan's workshop complete with fiery thunderbolts and the elemental ring of hammer on steel through air thick with brimstone. Atop a two-story unit, a lone welder is wreathed in green flame and plumes of smoke as he burns aluminum grate that will become the false decking in the Combat Information Center below his feet. In the dim light from behind, he looks as if he's on fire. A fitter down in MSA bangs out a slow dirge, the sound seeming to float in and out of the deserted steel hulks littering the floor.

The graveyard shift, a skeleton crew of about thirty on from 10 P.M. to 6:30 A.M., has a reputation as the haunt of the misfit and the troublemaker, the loner and the disturbed. As if to confirm this, a figure in welding leathers hurrying along the walkway at the back of the unit abruptly removes his hardhat and brings it to rest, curved side down, straight out in front of him in a single, robotic, movement. Still walking, he throws it down on the floor in front of him. The hat bounces off the concrete floor, does two perfect rotations, and returns like a yo-yo to his hands. He smiles inanely, plops it right back on his head, and keeps on going.

Twelve-fifteen, and Charlie Barnes's crew can be found spread out among the much-abused fake-woodgrain metal tables in AB's north end conference room. This first week of November, they eat chips and drink coffee or soda, waiting for Charlie to appear.

This weekly, time-honored tradition is known as the Safety Meeting, a half hour late Wednesday or Friday night dedicated to company business, a short talk on some aspect of work safety, and in general a good session of bullshitting and joking around and tall tales told in answer to questions like: Didja get yah de-ah yet?

Charlie's crew is evenly split between welders and fitters, and the two trades carry on a gentle rivalry over whose job is more physically strenuous, more dangerous, more unhealthy and dirty.

Welder or fitter, each job has its perils. This night Ted's eyes are red and wet. He keeps them half-closed, obviously in pain. You get flashed? Randy asks. Ted nods, explains that he was working around the array face, and with the pressure on, another welder right beside him. Every time he raised his hood to check his own work or grind, he got a blast from the other guy's torch. Like a sunburn—except on your eyes, Randy says, hurts like hell. You know about the tea bags, right? You wet 'em, wring 'em out, let 'em sit there. Ted nods. No, no, Mike says, joking, what you want to do is rub 'em real hard!

Mike has had his own close calls. Sometimes, he says, I feel like I'm playing the percentages—the longer you're here . . . It's a

Charlie Barnes's crew. Back row left to right: John "Gummy Bear" Collins,
Merle Witas, Ted Kramer, Eric Bowman, Matt Larochelle, Randy Fisher, Mike
Matthews. Front row standing left to right: Luke Arsenault, Jason "Spanky"
Cloutier, Gary Havelicheck (seated), and Charlie Barnes.

gamble. You get a little injury—hurt an eye, pinch a finger, but
you always wonder, when's that big one comin' at ya'? I've nearly
been flattened by couple tons of side shell, taken out by foundations. Fuck, I had staging fall out from under me forty foot up.
Been nearly blown up a couple a times inside a ship. I'd rather
work in here than out there, 'cuz in here, everything's out in the
open and you can see it coming. Out there, the guy who takes you
out might be five compartments away.

Mike proceeds to tell a story about when he was a tank
grinder, working inside the 50,000-gallon fuel tanks in the very
bottom of the ship, grinding out the rust after they'd been filled
with water to test for leaks. A welder had to fix some small part,
the only time he'd ever seen a welder down there, so Mike climbs
out of the tank to let him work. No sooner is he out than some
fool a couple decks away turns on the fill valve for the tank, flooding it with water in less than a minute. Mike pulls the welder out,
and the two of them take up hammers and go after the guy's head.
Mike shakes his head, laughing. For about two hours there, I was

a wild-eyed maniac running around wanting to know who the hell that ass was who turned that valve . . . until they sent me home for the night with pay.

Randy notices approvingly that Matt, the new welder, has shaved his beard into a Vandyke so that a respirator will fit around it. It's a point of pride with some welders not to shave, not to use the bright purple respirators, not to set up the sucker hoses that draw the noxious fumes of gases and burning metal away from the welder's face. He'll tell you how stupid that is, how dangerous, how they don't even know the effect of so many years of breathing that stuff.

You smokers there, Randy says, if you don't use a respirator, have a cigarette after you gouge. Tastes sweet, he says drily. I wear a respirator, Ted says, so I can still smoke. Even with a respirator, Randy continues, after you gouge, it tastes just a particular taste. Yeah, Ted says, makes menthol taste like regular.

This crew of eleven has a little of everything. It's got the very young—Matt Larochelle, a very quiet apprentice welder not yet twenty-two—with the older set represented by Steady Eddie and Gummy Bear. Steady Eddie, a welder in his sixties, doesn't hear too much and speaks less, while Gummy Bear has no teeth, which doesn't seem to stop him from talking at all.

The younger fitters could take lessons in cool from Ted. Dark-haired and whippet thin, he's got a great truck and a hot girl. He certainly knows how to fondle an unlit Marlboro languidly and manages to wear his short-billed, blue pin-striped welder's cap with a jauntiness that Jason "Spanky" Cloutier will probably never master.

Spanky—short and round and pale as his Lil' Rascals namesake—is the butt of a thousand jokes about his inexperience with women, and must hold the yard record for most blushes per hour. Before Charlie walks in, he and Mike are talking about Christmas.

My wife, Mike says, she's been mentioning how she wants a diamond for Christmas. I says, not a problem. You want one, I'll get you one. In return there's something I want—a frickin' Harley. She made me give up motorcycle riding years ago.

A good diamond costs, another mechanic asks, two thousand, three thousand bucks? A Hawg is what, eight, nine thousand?

So? Mike asks.

Yeah, Spanky asserts, a diamond, you know, you can only wear certain places but a Harley you can take anywhere!

Amid the smokers' coughs and scraping of chairs, Charlie opens the meeting, as he always does, by asking the guys what they need. Pen in hand, he scrambles to keep up with the flurry of shouts.

No kitty litter at number 2 door. No big bolts. Yeah, no big bolts. Where'd all the lug-all handles go? When are the new pulse-arc rigs gonna arrive? How come they're no assholes nearer 432 than two miles away, Merle complains. There's a big friggin' rack right by where the riggers store their shit, Luke tells him. Back and forth it goes, until Mike picks up on the mention of someone's seniority anniversary.

You know, at twenty years, he says, they used to give you the night off, with pay, and treat you to a banquet. I was in the last group that got to go. After our group, they shut it down.

I don't blame 'em! Ted snorts. They probably figured, Jesus Christ, as much as you ate, they'd go friggin' broke. Everyone cracks up, and Mike says, I hate you, too, Ted. Charlie is making quiet-down noises at the front, and finally asks, anything else?

He asks if everyone's seen the new paychecks and then reads the man-hour tallies from each unit, informing Eric and Merle that they've busted the budget on the deckhouse they were working. Job well done, says Mike, but does it have anything to do with the hundred and sixty feet of patches inside we hadda do?

The surveyors're doing the photogrammetry on it, Charlie explains patiently, and for some reason it isn't coming out. We're trying to talk them into going back to setting the unit up top, leveling it off, scribing it in place, then taking it off and burning it. But they don't really want to buy it because they've got all this money into the photogrammetry. Hopefully they'll get the bugs out. Soon! Charlie finishes, with an emphatic snort meant to signal the end of the discussion.

Photogrammetry is the surveyor's method of determining the dimensions of one unit in relation to how it must match up to its mate. A coded photograph of the top of one unit can be superimposed on another photo of the bottom of the unit that will sit on it, and determinations can be made about how much must be trimmed off and where to attain the necessary between-decks height.

Steel shrinks when it is welded, so much so that no two DDG-51s are ever exactly the same length, width, or height. Through long experience the engineers know about how much the steel will shrink and where, and so plan for it. When it comes time to mate up levels of a unit, to do all the long butt welds of shell plate to shell plate and deckplate to deckplate and join all the structural members to each other where they meet, in the best-case scenario there is a "tight butt" with just enough "fat" to account for the welding shrinkage. Too much fat, and steel must be trimmed.

The opposite case—a gap—means rework time. This is because the Navy will not allow BIW to fill a gap beyond a certain width with built-up weld. Instead, they require the shipyard to put in a patch. If a section of bulkhead is half an inch short, the fitters must remove all the steel—panel and stiffeners—within a foot of the bottom edge, and make up and weld in a new piece large enough to eliminate the gap. It's expensive and time-consuming, Mike says, and basically not a good thing.

It is a point of pride with this group that, barring unforeseen disasters, they most often deliver their assembly units on time and on budget. Patches, specifically on whose tab the extra man-hours go, are thus a heated point of contention. As are stolen tools, which is why the group keeps its own locked toolboxes for those things apt to go awalkin'. They expect Charlie to stand up for them, to shield them from as much of the corporate b.s. as possible, to resolve workplace disputes fairly and hopefully without resorting to grievance committees. In return, they'll stand up for Charlie and work their asses off for him. If he can consistently get them the tools and materials they need, keep the interfering white hats at bay, and make sure safety issues are dealt with instantaneously, he's a local hero and his name will go down among them as a stand-up guy.

The opposite side of this coin Mike Matthews pointed out one evening earlier that week, walking past an erection unit two decks tall. On the top deck, a half-dozen 15-ton weights rested, the sign of a bad crown that somehow had escaped notice until this late stage in assembly. Sometimes, Mike said, those are our last resort, those weights, to try to drive a unit down. I've seen 'em have as many as ten, twelve of 'em on top of a unit stacked two high. Mike shakes his head at the sight. The day-shift supervisor who's on that doesn't really know that unit, he says. Another day-shift boss sent his pride and joy down there to set him on the straight track today. So we're gonna be watchin' him like a hawk. He may not stay in charge too long.

Charlie is in no danger of being sandbagged by his crew, who clearly respect him to the point of treating him exactly like they treat each other. Oh, they'll bitch and moan and make faces behind his back when he scolds one to wear his safety glasses out on the deckplate or to do something in a less dangerous way (BIW is in the middle of a safety campaign). Yet he'd be asked to go deer hunting with some of the guys without, one mechanic chuckles, having to worry about being shot in the back, unlike some white hats around AB.

Back in the meeting room, Charlie's intoning, in the midst of a fire safety lecture, one of BIW's most often repeated rules. Before performing any hot work, he says, always know what's on the opposite side of the bulkhead or deck before you strike that arc or light that torch.

The mechanics sprawl back, reading newspapers, muttering about weekend overtime and union business, and generally paying little attention until Mike asks, out of the blue, You guys know where the hell to run if the bulding starts burning down? I certainly hope so! says Charlie. They obviously don't, confusion reigns, evac routes are discussed, and the men sit up and look a little more alert until 12:40, when the five-minute-to-shift-change whistle blows.

Out in the yard, the wind coming off the river is surprisingly mild, the ground bare. Maine is having a gentle early winter, and the guys are complaining about how hard it is to track a wounded deer when there's no snow on the ground to highlight the blood. They wait patiently by the time card machine near a side door, talking about a white buck someone shot up north the weekend before. My granddaddy told me that was the soul of a hunter lost in the woods, one says, and then the whistle goes, and they begin to file out into the blackness of the vast unlit tarmac between buildings and ways. The crane arms pivot noiselessly overhead, their spotlights illuminating the two hulls abuilding. The whine of a grinder lost in the tangle of staging down near the river's edge mingles with the last echoes of the whistle. The men quickly disperse, disappearing into the late-night darkness. Eric pauses to light up one of the few cigarettes he allows himself each day. He and the rest of Charlie's crew will be back here Saturday noon, picking up where they left off. High spots 'n low spots, he says, his smile white in the gloom, low spots 'n high spots.

CRANE

The riggers, they're the bosses. I just pull levers.

—*Jason "Buffy" Knight, Operator, Crane #11*

On the approach to Bath from any direction, the first thing to catch the visitor's eye is invariably the towering arms of the two massive riverside cranes that dominate the skyline. Decked out with a Christmas tree atop one upraised boom at the holidays, or flying an American flag as big as a house from a special hanging rig on the Fourth of July, the cranes are as much a part of the town as the yard and, indeed, have come to symbolize it just as much as the Bath-built sailing ship that is one of the town's official emblems.

Crane #15, running on five hundred feet of railroad track parallel to A-Ways, is technically called a *level-luffing* crane, with a nearly three-hundred-foot boom and a 330-ton lifting capacity. It was bought for $7.6 million from the American Hoist and Derrick Company and put up in 1987. Its surprisingly delicate white-painted boom, fabricated of welded beams and braces, lets the frequent high winds up the river blow through. The cab sits atop four cantilevered legs nearly eight stories above the ground, a squat, glass-fronted cabin with the boom jutting up at one end and a stack of seven 72-ton counterweights at its rear.

Even more massive, rising four hundred feet above the river, is Crane #11, running on rails between B-Ways and C-Ways, its orange- and white-painted boom faced with steel plate, giving it a more substantial presence, even though its lifting capacity, at 220 tons, is only two-thirds of its more fragile seeming neighbor. Fabricated by BIW from a Japanese design in 1973, it has long been the workhorse of the yard, even though its boom design makes it more sensitive to wind.

BIW's need for such large cranes has only come about over time with its increasing sophistication in preassembling units, in erecting units on the ways that are more and more complete—and thus heavier. The heaviest units BIW erects today are the main engine room units, which at 330 tons are too much for even Crane #15, given that a crane's lifting capacity decreases as the boom moves down towards the horizontal. To lift one of these units both 11 and 15 will work together, one on each side.

BIW has five smaller-capacity cranes rated from 25 to 90 tons along its piers, but the big cranes are absolutely necessary for unit erection. They are also necessary to the functioning of AB, since heavy pre-outfitted units that start out inverted must at some point be flipped for further right-side-up work and when it comes time to join several decks together to form erection units. Aside from the awkwardness of flipping units inside the building—there's just not always the necessary ceiling height—the inside cranes are overwhelmed by the ever increasing weight of units being outfitted to greater and greater degrees. The way-side cranes also move tons and tons of equipment, tools, and material from the ground up onto, or even into, the ship as it is erected.

Five A.M. on a brisk March morning, still dark, a glance down during the long climb up to the cab of #11 reveals little more than pools of light thrown haphazardly by the sodium vapor lamps scattered around the yard. You can make out a single radar mast on the ways, the dark hulk of AB and the Aluminum Shop at the back of the yard, and in between, two keel units up on the tarmac above the ways, spilling sparks and noise as a gouger trims steel. As you lose count of the number of flights of stairs you've climbed, the legs start to burn, the lungs pull in gulps of salt-laden river air. Finally the bottom deckhouse door opens onto a humming machinery room. You circle a hub, climb another staircase, a ladder, go out another door onto a catwalk leading to Jason "Buffy" Knight's domain, an eight-foot by eight-foot cab perched out on the front right side at the base of the boom. If you're afraid of heights, now's really not the time to look down, or around for that matter, because even in the half light you realize that you're up higher than the Carlton Bridge, the whole yard and town beyond spread out before you, so high up, in fact, that you can nearly see over the high ridge behind Bath into the valley behind it.

Fitted and welded up inverted as in Figure 17, an assembly unit now gets flipped over outside AB. Facing away from us is the unit's flat, deckplate side. While the crane dangles the unit straight up and down, the two riggers will detach the slings on the bottom, manually rotate the unit 180 degrees, then reattach the slings. The left chainfall tightens, bringing the unit horizontal, then both chainfalls lower the unit, flat side up, onto the low, flat transporter glimpsed in the background.

Cranes #11 and #15, one on each side, lower the *Donald Cook*'s 330-ton main engine room unit onto the ways. Behind it, the first unit to be erected already sits on centerline blocks, stubby inner shores, and angled side supports.

Inside the cab Buffy's got an armchair center-front, almost against the floor-to-ceiling glass wall that looks down, this morning, on the few keel units sitting on the ways, all that has been erected so far of Hull 468, the USS *Howard*. The controls for so massive and complex a piece of machinery are astoundingly simple. His right arm rests on a raised platform at elbow level, his hand gently moving a lever forward or back, raising or lowering the load at the end of the boom. On the right is also a lift gauge whose needle ticks off the weight of the load in tons. A Ziploc bag full of crackers is perched up against the weight gauge, Buffy's all-important midmorning snack.

With one foot Buffy controls the TRANSMIT button on the indispensable two-way radio that connects him to the riggers working below. The microphone is mounted on a metal rod off to the left, out of the critical window of space directly in front of him where the view of the load—and the men around it—will always be.

His left arm rests on a similar platform, which has two controls and one further measuring instrument. One lever moves the boom up or down, another directs the electric motors that turn the train wheels on which is mounted the entire crane structure, moving—"traveling," the riggers say—the whole crane backward and forward along the tracks that parallel the ways. Both cranes have about five hundred feet of track, and this gives them a huge reach in terms of the area of the shipyard they can service. This is critical for the heaviest lifts, around 300 tons, that require *both* cranes, one lifting the port, the other the starboard side of a unit.

The tremendous reach of both the big cranes is also attested to by the signs above Buffy's head. "Caution: Number 11 crane boom can hit Number 15 crane boom," says one. "Caution: Number 5 crane boom can hit Number 11 crane boom and machinery house." Which has, rarely, happened. These cranes are extraordinarily powerful pieces of equipment. All riggers have their horror stories, usually involving what happens when a crane tries to pick up something that's still attached to the ground—or a building, in which case, depending on the relative strength of chain, shackle, or lifting pad, either the unit gets bent and wracked or a corner of the building is pulled right out of the concrete. Which has also, rarely, happened.

Mostly, however, things do not go wrong, and they don't go wrong because crane operators do not get a second chance. Buffy may chuckle and giggle and joke a mile a minute with the riggers, but his eyes do not stray from the rigging crew a hundred-odd feet below him, often working around and under the 100 tons of steel he's in control of. If his eyes do stray, it's because he's stealing a glance at the wind sock on top of the Aluminum Shop, or because he saw a stray piece of canvas flap, or because the unit he's "flying"—moving through the air—is acting strangely. In which case the next place he'll look is at the instrument to his left.

That instrument is a wind speed indicator. Wind can blow up in this part of the world in a matter of minutes, and a significant gust will push several hundred tons of steel around in the air like so much cardboard.

The riggers, Buffy says, they're the bosses. I just pull levers. The only thing I have power over is saying I can't run the crane on account of the wind, or weather-wise, anything I don't feel comfortable with. Usually, shutdown on this crane is 36 miles per hour of wind. Depending on what we are doing, I can decline the lift. Some things—say, something we lift is too light, too gormy—the wind'll catch it, like plywood, or too awkward, lots of sail area, bulkheads. Buffy shakes his head.

This morning Buffy's got the crane positioned far forward at the head of the ways, with the chain and wire-rope lifting arrangement dangling from his boom over the 3340 unit, a flat, single-deck stern unit that sits up off the ground on massive steel pillars called *plats* or *pylons* in front of the Aluminum Shop next to AB. He looks down on a handful of orange-hatted riggers who've spent the last half hour hooking up shackles to connect the slings to the lifting pads welded to the corners of the unit. This is one of the few single-deck units that go out to the ways for erection as a single unit; and at almost fifty feet long by fifty-four feet wide by nine feet tall, it is exactly what Buffy means when he says "gormy." It's light and, without a huge amount of structural steel, wobbly even in a medium breeze. The day before, with gusts up to 50-plus mph, it would have been flopping around in the wind.

Richard Douglas and Mike Shaffer are working the lift, sending a constant stream of requests to Buffy, who, more than a hundred feet above them, can't always see them as they disappear under or behind the unit. Along with Burt Boisvert, Pete Watson, and David Huggins, they're looking to make sure first, that the unit is free of any attachment to plats or ground, and second, that the slings are positioned such that it will rise evenly from all four corners.

This is where BIW's "real estate" problems can easily be seen. There are other units, pieces of machinery and equipment, steel scrap and debris just a few feet behind and beside this one, all crammed into the only bit of open ground beside the roadway that's got to be kept clear for the transporter that maneuvers the units in and out of AB. In addition, once airborne, the unit has to travel over a set of temporary trailer offices, which must be evacuated for every lift, and around or over the forest of staging that lines the ways port and starboard. This is why BIW tries to do most unit erections at 5 A.M., when the yard is at its emptiest and other calls on the cranes' time are at a minimum.

On the ground all is ready, and Richard gives the order over the radio. *Okay, Buffy, up easy . . . hold . . . They got a pipe they're gonna take off here first.* Minutes pass before Richard keys his mike again. *Now you can nudge her up easy. Holdin' about 84-ton now,* Buffy tells them. Richard: *Comin' up easy, Buffy. Comin' up,* Buffy replies. *Hey, Buff,* Mike Shaffer says, *travel east just a touch.* Buffy moves a lever on his left a fraction of an inch until Mike comes back with, *All right, hold that! You all right, Peter?* he continues, asking the status at the other corner. *Up a little more, Buff.*

The unit comes suddenly free of the ground, sidles backwards into a pile of debris beside it with a shriek of steel against steel audible up in the crane cab, this followed by a secondary crunch of metal. The whole crane shakes and shudders as it lifts the full weight of the unit. Mike continues to direct. *Travel east just a touch, Buffy,* he calls. *All right, hold that. . . . Bump her east a little more . . . comin' up easy, coming right up . . . you're all free.*

What you wanna do, Richard tells Buffy, *[is] take it up high enough to clear the staging and swing it south, then take it down through. Yeah,* Buffy replies, *I gotta boom up a little bit, too. How much you got for weight,* Mike asks. Buffy leans over and reads the scale on his left. *'Bout 92 ton. When you're ready, go ahead and swing it soft as you take it easy.* Now with the unit ten feet off the ground, two riggers man the ropes they have strung from opposite corners in order to spin it left or right to guide it through the maze of scaffolding and other obstructions should the wind begin to push it around.

How did it come off the beams, Mike? Pretty good? This is Burt, another rigger who's waiting on the already erected unit on the ways to which the 3340 will be mated. *No,* Mike replies unhappily, *not too good.* With the unit about twenty feet off the ground, Buffy stops lifting, letting the unit, which is bobbing up and down and swinging infinitesimally left and right, settle down. Up in the crane cab you can feel every sickening bob and wobble.

Buffy shifts a lever to his left, and the crane, with the unit following, travels down the ways towards its place at the stern of the ship. A bell rings, letting people on the ground know that the crane has begun to travel. Buffy gives two blasts on the air horn and hits a siren whose wail rises and falls irregularly. *So's people don't get used to just one sound,* Buffy says. Riggers appear atop the unit on which Buffy is setting its mate, passing the guide ropes into the hands of more riggers waiting. *Okay,* Richard is saying, *that oughta be far enough south if you like, Buff.* Mike is visible at the forward end of the unit, flapping his hand. *Move it this way a little bit more, Buff. Is it gonna fit on top of this one?* Richard jokes.

Start bringing it down a little bit? Buffy asks. *Yeah-uh,* Mike replies. *Aft a bit more, too,* Burt adds. *Yeah, about four feet,* Richard says. *Hold your travel, Buff. Swing it north some, we gotta go north quite a bit.*

As the 3340 unit drops down closer and closer to the deck to which it will be welded, the riggers' problems multiply. Buffy is looking down on a flat deck from whose underside hangs all manner of bulkheads, together with the unit's pre-outfitted pipe and duct runs, and pieces of equipment. All these things hanging down must, as the unit descends, fit over all of the other things sticking up—more pipes and machinery—from the deck below. And with even the littlest miscalculation, 90 tons of steel will demolish any interior obstruction, causing expensive rework.

This is why Mike spends about twenty minutes getting Buffy to lower the unit about six inches and move it forward about a foot. He's making sure the 3340 is lined up fore and aft with the centerline of the unit below so that, when lowered onto temporary blocks, it touches down all four corners at the same time. Otherwise, it being a "gormy" unit, it might be bent or wracked should all its weight land on one spot.

Mike has stopped everything, and blue light leaks out of a door-sized oval at the aft end as a welder cuts a piece of pipe that is sticking up too far from the floor. Richard and Pete each man a special block and tackle rig, one to port, one to starboard, that allows them to raise or lower the forward end of the unit manually, in minute increments. They start working the chains, lowering it, until Mike stops them. *Hold up there, Richard,* he calls. *We got a problem. Buffy, I need you to come up a bit. Now I need you to travel ahead about an inch or two . . . that's good.*

Yes, Buffy can caress that lever just enough to rotate the crane's wheels an inch or two, and he does it without thinking. This is something you can get used to and judge, he says simply, and the height doesn't bother me. You get to know your riggers well, too well!

Okay, Buff, Mike is saying, *now come down some, little more easy. Touchin' over there, Pete? The aft corner's touchin',* Pete replies. *Go ahead and pick 'er up a little, Pete, so we're clear. Okay, Buff, come down a little. Yeah,* Mike says a few moments later, his voice a bit frustrated, *this thing must be wracked. It's touchin' in two corners, and the other two opposite aren't.*

So it goes, inch by inch, for the next hour and a half until it's time for the next step. Eventually Mike and Richard and the others will get the unit lined up on the centerline of the already erected unit below it, and setting on blocks at the corners. Then the surveyors come in and, using photogrammetry like in AB, determine where extra stock steel on the bottom edges has to be cut off. They'll actually scribe a line into the steel where the shell plating of the hull needs to be cut down or where bulkheads are too high and need to be trimmed.

Then Buffy will pick up the whole unit, fly it over the ways back to the tarmac, and set it back down on the plats for the second-shift welders to work it. Tomorrow or the next day or next week, Buffy will fly it back, all of the edges where it will butt up against its mate cut neat and ground to a high shine, ready for the final weld-up.

And so it will go, unit by unit, starting in the middle of the ship and working fore and aft and always up, deck by deck, over the approximately six months it takes to put each ship together. Buffy will be there every workday, high up in his perch, chuckling and eating crackers, gassing with the riggers, his eyes glued to them as they work below, always mindful of the fact that he holds their lives in his hands. I've been up in the cranes eight years . . . and I'm still training! he jokes. The riggers know differently; they'd refuse to work, period, with someone they didn't trust.

DRIVING
THE WEDGES

Don't tell management this, but I'd work here for free.

—Tim Vear, Electrician on Aegis control rooms

It is midnight when Dean Atkinson pulls into the half-empty lot beside the shipyard one brisk evening the first week of May. Aside from the intermittent muffled clang of steel on steel from grave-yard-shift welders and pipefitters in AB and the occasional truck passing over the Carlton Bridge, the night is quiet. The sky is clear, brilliant with stars as only a maritime sky can be, and the faint breeze off the Kennebec River cool and salty. Good weather now, but the networks are predicting a cold rain for later in the morning when the ship is scheduled to go off.

Dean makes to flash his ID at the guard behind his glass walls, but the guard just waves him in, sticking his head out the door to offer a hello and a wish that the weather will hold through the afternoon. Dean passes through the narrow entryway into the yard proper, settling his hardhat on his head. First he heads to his office, up three flights of rickety wooden exterior stairs to one of six dilapidated construction trailers stacked two to a level on a makeshift frame of welded I-beams just off A-Ways.

Inside, the linoleum floors are cracked and faded, plywood subflooring peeping through the rips. The air smells faintly of burnt coffee and sweat. Steel desks line one wall, their surfaces rough with grit blown in from the hulls blasted and painted less than one hundred yards away. The walls are covered with photo-copied diagrams, a side view of the ship ruled into the forty-odd modular units out of which it is assembled, scattered details and exploded views of single units, the October erection calendar, various schedules, and a dozen yellowing corporate memos with titles like "Smoking in the Workplace," and "Sexual Harrassment on the Job."

In his office Dean grabs the Motorola and clips it to his belt. The radio will be squawking at him—or he'll be squawking into it—almost constantly for the next twelve hours. He looks for Chris Medeiros and Tom Niles, his two senior supervisors, wanting to check in to see that their teams are ready and that last-minute work has been seen to by the mechanics just going off shift.

He's tired, considering he was here all day and had only a few hours of rest and some food at home earlier in the evening. He calls up Tom and Chris on the radio, telling them to meet him in the office, then kicks his lean frame back into a beat-up swivel chair. At age forty he has a mane of thick shaggy hair gone completely gray—from too many launches, he jokes—over blue eyes deepset into a narrow face. A ruddy mustache clashes with the gray, and if it weren't for the constant smile, a stranger might think him ferocious.

A Mutt and Jeff pair enter—Tom beefy, baby-faced, and clean-shaven in contrast to Chris's wiry build and slight features, his dark hair and mustache. They go over some last-minute details of Dean's eighty-five-item "Hit List," all things that must be attended to before the launch can take place. As they talk, the room is filled with the soft echoes of regional Maine: . . . movin' this ovuh heah? Shoulda been done yestiddy! Shoah, and he'll be down to the south stores, yeah? They are respectful, listening intently to each other, three guys who have worked together for more than fifteen years until their launch crew runs like a finely tuned machine.

Dean is tired but he's pumped. A little less than a year ago, on July 9, 1996, at about 4 A.M., he and his crews were clambering around on the empty ways with the first completed keel unit of this ship—a 100-ton, half-moon chunk of steel plate and beam, pipe, wire, and equipment—flying thirty feet overhead, swaying

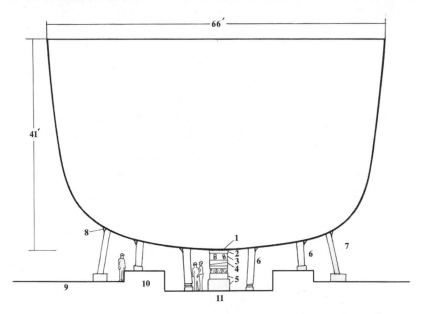

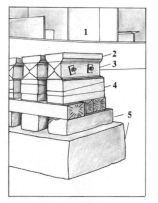

Ship on the ways, on shore support, from initial unit erection to just before launch. From stem to stern under the ship's keel (1) run stacks of centerline blocks, each of which consists of (moving from keel down to ground ways): (2) fitting block, (3) quarterblock, (4) marrying wedge, (5) concrete keel blocks. There are three sets of shores from the centerline out: (6) stubby inner shores and (7) angled side supports. All shores would kick out under pressure but for (8) the shoring clips welded to the hull. The two parts of the ways are (9) and (11) the building ways, and (10) the launching ways.

ever so slightly in a stiff breeze. As Dean watched, the orange-hatted riggers, whose specialty is moving things large and small, directed by radio the operators of the two largest cranes, Crane #11 and Crane #15, in the unit's gentle descent and laying-in on *shore support,* a bed of stubby wooden shores and pyramids of blocks underneath the keel, and side shores angled in against the side of the hull above the water's edge of A-Ways.

Over the next six months more units were laid in, alternating fore and aft fom the middle until the entire keel was in place, after which all of the upper deck and deckhouse units were laid in and welded up. Four months ago, with the ship nearly together, the same crews began the long and critically detailed process of constructing the wooden launch cradle according to principles as old as boatbuilding itself. The idea is a simple one, beginning with the *ways*, a ramp sloping slightly from shore to water. (See page 125.)

The mechanics who work the ways launching ships informally use different names for different parts of the ways, even though a naval architect would correctly refer to the whole structure as simply the "construction ways." There are the *launching* ways, which appear as two wide elevated parallel tracks, between them a sunken space, like the grease pit in a garage. The sunken space together with the areas outside to the left and right of the launching ways are the *building* ways, working space usually crowded with tools, parts, scaffolding, and everything else necessary for ongoing work on the ship. Resting on the launching ways is a cradle that must support the weight of the ship at completion. At launch, ship and cradle slide together down the launching ways into the water to a point where their combined buoyancy floats the ship off.

Surprisingly, there are most probably many more similarities than differences in launching the *Virginie Sagadahoc,* the first ship built in America just down the river in Phippsburg in 1607, and the *Donald Cook,* scheduled to go off the ways in a matter of hours, 390 years later. This despite the fact that the *Cook* is two thousand times as heavy, five times as high, and ten times as long. Some of the first inhabitants of the town of Bath were English fisherman and shipbuilders who brought hundreds of years of shipbuilding tradition across the Atlantic with them.

As Erik Hansen, the tall and ascetic-looking Dane in charge of Hull Engineering explains, you don't sit down and say, now I'm going to design a launching arrangement to launch a ship. They launched Viking ships a thousand years ago without the benefit of engineers, he says with a laugh. These processes are being codified and extended at the same time.

In other words, BIW's ship-launching know-how owes more to the historical antecedents of all those Kennebec River shipbuilders who came before than it does to some out-of-the-blue modern-day technological innovation. Elsewhere in the yard, whether it's bending aluminum lifeboat davots or assembling a length of pipe with thirty-six deviations in a twelve-foot run, the same credo holds: Look at what they did before you and learn from your own mistakes.

The cradle, whose design Erik's group is responsible for and which Dean, Chris, and Tom's ways crews built, is a good example of this kind of evolutionary knowledge. Almost ten years ago Hansen first looked over the *scantlings*—the raw dimensional descriptions—of the *Arleigh Burke*, the first in the DDG-51 class of destroyers of which the *Donald Cook* is Bath's fourteenth. Considering the thicknesses of the plates and beams, the shape of the hull, and the weight of the ship, he and his Weight Control group figured the rough launch weight based on how much of the ship would actually be completed at the time of launch. This then determines the size and shape of the cradle.

The concrete launching ways are like two elevated ten-foot-wide tracks running down into the river. First the mechanics lay a bed of timbers, called *sliders*, in sections connected by heavy steel strapping on top of the greased concrete. On the outside of each launching way is a raised lip of steel, the *ribbon*, which keeps the sliders from veering left or right as they travel down into the water. On top of the sliders, wooden blocking called *solid packing* is piled up to the hull, its height greatest near the bow and stern, where the hull arcs up and away from the ways. Bolted every ten feet to the outside of the packing is a vertical oak plank five inches thick and ten inches wide, its height varying with the height of the packing at any one point. This is the *hutchet*.

To keep the weight of the ship from pushing the piled blocking out, tie-rods run through the hutchet and blocking on each side every ten feet along the ship's length, connecting underneath the hull with cables. Where the packing is highest, three sets of tie-rod and cable may run from port to starboard hutchet to hutchet. Where the packing is very low, a single tie-rod and cable may run between hutchets. To counterbalance the inward pull of the tie-rod and cable, long timbers—*spreaders*—are forced into place at intervals between sections of port and starboard sliders.

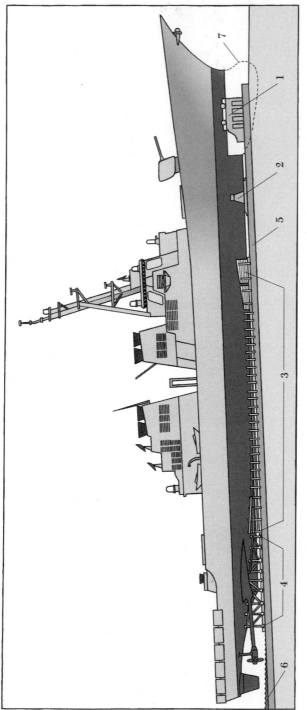

Side view of the *Donald Cook* on the ways at launch supported by (1) fore poppet, (2) saddle, (3) solid packing, and (4) stern poppet, which together with the sliders underneath these structures make up the cradle. At launch, ship and cradle slide down (5) the launching ways until they enter (6) the river at the end of the ways. The shape of the sonar dome (7) at the bow to be installed after launch with the ship in drydock is represented by a dotted line.

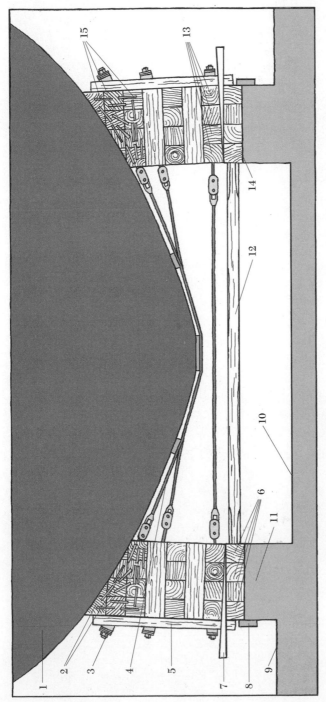

The ship, head-on view, on the cradle just before launch, with all shore support removed. (1) The keel, (2) fitting blocks, (3) tie-rod penetrating packing to attach to (4) tie-wire, (5) oak hutchet, (6) sliders, (7) oak wedge, (8) steel ribbon, (9) and (10) building ways, (11) launching ways, (12) spruce spreader, (13) canting blocks, (14) layer of grease on top of layer of wax between way and sliders, (15) Peter Marshall's timber dogs used to staple packing together.

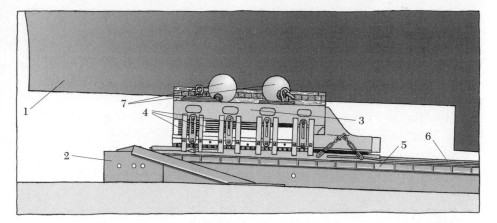

Side view of the fore poppet at the very head of the ways. (1) Ship's bow, (2) head of the ways, (3) fore poppet made up of steel collar, and (4) layered pine and plywood crushing arrangement, (5) ribbon bolted to outer edge of ways, (6) sliders on which fore poppet rests, and (7) marker buoys for locating the poppet in the river after it falls away when the ship is in the water.

Supporting the bow and stern during launch is more complicated because the shape of the hull changes so radically in these areas. Near the stern, the shape of the hull changes from where the prop shafts exit the hull down to the props themselves under the flat and much shallower fantail. The stern is so heavy, and also so high at this point relative to the ways, that steel replaces wood between sliders and the hull—in this case, steel beams with crisscross supports, with a thin layer of fitting blocks where the supports meet the hull. This structure is called the *stern poppet,* and it is triced tightly to the ship in order to keep it in place during the rigors of launch.

Amidships, the bottom is a nearly uniform, flattened U shape while, just at the arch of the bow, it soars up in a steeply angled V whose sharp edge and abrupt climb make it impossible to fit blocks to it, never mind the enormous cost in labor and materials of piling lumber to a height of twenty feet. Also, the violent nature of stern launching on ways causes engineers to worry about piling timber any higher than they have to. As the ship moves down the ways stern first into the water, there will come a point—called *pivot*—when the stern will begin to lift. At pivot, 20 percent of the weight of the ship will be thrown forward onto this short section of hull still moving down the ways.

To support the bow during this critical moment of the launch, the yard constructs a *fore poppet,* a wide steel platform whose top edge butts up against the bow and whose bottom rests on twenty-four inches of soft pine blocking alternating with layers of plywood. This is called the *crushing arrangement.* Under the force of 9 tons per square foot during pivot, the plywood and soft pine compress, absorbing and dispersing a load that might otherwise buckle the bow.

From the fore poppet to where the keel flattens out, there is one very steep and thus problematic stretch of about thirty feet where wooden blocks are also too impractical for support. To stiffen the bow and keep the sliders in place (they want to buckle upward), Erik uses a *saddle,* a hollow, U-shaped band of steel that wraps around the belly of the ship, with blocking from its upright arms down to the sliders on each side of the hull.

These five parts of the cradle—sliders, solid packing, the two poppets, and the saddle—work together during launch to form a dynamic system. It is dynamic not only in that cradle and ship move together down the ways, but also owing to the physical properties of its one-foot by one-foot wooden timbers, which have enough play to absorb the enormous—and unequally distributed—weight of the ship without breaking.

This ability to compensate for unequal and locally shifting loads is actually built into the composition of the blocking between sliders and hull. On top of the sliders sit foot-square, soft white pine *canting blocks*—they cant, or tilt in, to meet the slope of the hull. Between sliders and canting blocks hundreds of oak wedges are inserted, and it is their sloped faces that make the square canting blocks tilt. On top of the canting block sits a *fitting block,* whose upper face has been carefully milled down to fit the hull exactly.

Every wedge—and there are more than two hundred of them up and down each side of the hull—can be thought of as distributing the ship's weight along the entire four hundred and twenty feet of port and starboard sliders from bow to stern. When the wedges are driven in towards the center of the ship, they force the blocking up however far it needs to go to embrace the hull. They provide a means to compensate for irregularities in the surface of the hull, in the blocking, and even for minute imperfections in the ways themselves. If a section of blocking is low, for example, the wedge can be driven in further, forcing it up to the hull.

The final constituent of the launch cradle is what makes it slide. Raw wood just doesn't slide very well over rough concrete, so the tops of the launching ways are prepared with a layer of hot wax before the first slider is laid. When the wax has hardened, bright orange grease is troweled over that an inch deep. To keep the weight of the cradle off the grease until just before launch, grease irons—steel strips on which the sliders will rest—are laid at right angles into it every ten feet. This wax/grease combination, the hardend wax making up the *basecoat* and the wax the *slipcoat,* becomes adequately slick only under sufficient pressure. It is tricky stuff, as Erik or Dean will tell you, vulnerable to the vagaries of temperature and moisture.

Back in the trailer Dean pulls on a jacket and heads out to the ways for a final inspection, the last cleanup from his infamous Hit List. Two weeks earlier he was out here at three in the morning, pen and notebook in hand. He likes to come when there is no one around, no one to disturb him with questions or problems. He starts at the stern and works his way forward, smoking one Old Gold after another, looking, touching, crawling through the timbers to inspect every inch of the cradle.

About the worst thing, he says, that could happen during a launch is some obstruction between cradle and ways that interferes with the free movement of the ship. A low hutchet, say, one of the oak five-by-tens on the outside of the cradle and through which the tie-rods are strung.

You got what they call a ribbon on the outside of the sliders, Dean continues, acts like a track for the ship. If you have a low hutchet when you launch, you'd obviously break it [against the ribbon] and create a series of things. . . . Well, he'd rather not think about what would happen.

For instance, that the downward motion of the ship, which

cannot be stopped once started, would wrench the hutchet off the cradle, pull down that particular section of solid packing as well as all of the packing on the opposite side with no tie-rod and cable to hold it in place. This, in turn, might cause one section of slider to buckle, putting enormous pressure on the sections immediately fore and aft of it as well as creating a massive obstruction through which the rest of the cradle, saddle, and poppet would have to pass cleanly and without being destroyed for a launch.

And cradles are expensive, about a million dollars. So expensive that they are specially constructed for recovery. Two thick steel cables, each with one end attached to one side of the cradle and the other anchored in the ways, reel out as the ship launches. Dean looks very carefully to make sure that these, the *snubbing wires,* have been strung and wired in correctly.

So when the ship's out there [off the ways into the river] four hundred some-odd feet, Tom Niles explains, the cradle literally stops because it's secured to the ways, and the ship keeps on going. That's what's supposed to happen. On the destroyer *John S. McCain,* we pulled the snubbing wires right out of the ways and the cradle stayed underneath the ship. That was a lot of energy going on in a little bit too close proximity of where we were standing. Nothing happened but . . . Well, you're talking two-and-one-eighth-inch wire, and the pressure of the surge onto those is about 100 ton. When you're twenty or thirty feet from them, that's closer than you want to be. Too close.

Other things go wrong, too. On the *Curtis Wilbur,* the destroyer launched in 1992 just before the *McCain,* the ship went off beautifully, no problems at pivot. But then, at snub, the cradle separated, actually broke apart, just below the saddle, and the ship took the rest off with it. It took three 1000-horsepower tugs to overcome the buoyant force of all that wood against the hull to pull the remaining sections out from under the hull.

Normally, at full extension, the velocity of the ship overcomes the upward, buoyant force of a couple of hundred thousand pounds of wood pushing up against the hull, yanking the cradle out from under the ship. Since every single piece is connected by rope stapled to its outside, the cradle stays together, floating just off the way-end to be corralled by a tug and towed away for salvage.

■ ■ ■

One mythical piece of launch history that floats around BIW like
so much driftwood in the river is known generically as the "John
Wayne Story," and it illustrates just how little it takes to turn a brilliant start into a million-dollar disaster. Dean was there, as was
Erik, but the more complete version is known only to a few people.
Ed Moll, who retired in 1997 as the chief of Hull Engineering and
was responsible for more than a hundred launches in his time, led
the forensic investigation of the events, and he knows it best.

The most common problem in launching, he begins, is either
a very slow start, or—we did have one in my experience looked
like it never *would* start. The *Oliver Hazard Perry*, the lead ship
of the FFG-7 class [the first of twenty-five guided missile frigates
that BIW built for the Navy starting in 1977]. It's captured on
video. I not only saw it personally, I also saw it at the beginning of
the John Wayne one-hour special because, coincidentally, he had
been invited to the launching by one of the directors of
Congoleum [which owned BIW at the time].

So, Moll continues, John Wayne was seated on the launching
platform, normally there's only about twenty people there. The
sponsor, the escort, and other distinguished individuals. *And
everything had been done*—bottle of champagne broken, trigger
down. Summertime launch, thousands of people, and Bill
Haggett, BIW's then-CEO, sort of leading the whole ceremony.
He'd seen enough launchings to know that not every [ship] starts
promptly. He's experienced at making small talk for whatever
time is necessary, which he did to a point. Then he ran out of
things to say, and the ship had not started, and the crowd was
totally silent.

The very first set of sliders has a steel head so that, should a ship
not move at launch, the crew can help it get unstuck by applying
pressure to jar it free. There are motor-driven hydraulic jacks
already set up between the head sliders and a solid steel barrier to
do just that, should the need arise. The launch crew, Ed continues,
is jacking onto the steel headpiece of the lead slide. The problem is
that, with all that elasticity in that wood, it takes a significant time to

edge the cradle along. And that's in process. But we've done a couple of these cycles, and the thing still hasn't started!

And this is what makes Maine truly unique—there's always some guy with a smart comment. Somebody in the back of this summertime crowd cupped a hand around his mouth and said, *Hey, Duke! Give 'em a hand!*

None of us [engineers] knew that John Wayne was there. Sure enough, John stands up in the back of this launching platform, and, with his great theatrical presence—I mean, he is definitely a big man, even from the distance I was watching, he looked big to me. He walked down one side, crossed in front and acknowledged the crowd just like he was in front of a camera, then came down the center aisle to the stem of the ship.

I've looked very closely [at the video] to see what he did, and he never actually touches the ship. It's amazing. He just sort of leans in that direction. And all of a sudden, that ship started. The crowd—a couple thousand people—just roared this great approval. John had this young female press agent with him. She threw out to this crowd these printed signature cards. Hundreds and hundreds of them, threw them out to the crowd.

Moll is doubled over with laughter at this point, hooting at the absurdity of the situation.

So, now, he continues, this is when you have to believe in good fortune. At the reception, people said that was the best darn launch they'd ever seen! The next week, the sales manager of U.S. Steel for the Northeast called me, he said he saw this launching on the six o'clock news and did that really happen? I said, yes, it sure did. And he says, well, you have the best damn PR I've ever seen.

After all the hoopla comes a call from the supervisor of Shipbuilding [the Navy's all-powerful on-the-spot overseer] to me asking what the heck went wrong there.

Our slipcoat was totally contaminated by sandblasting grit! Somebody made the judgment that we could launch a ship over that lousy, no-good grease. We had had a lot of pressure from the Navy to accelerate the date of the launch for whatever political purposes; they wanted to get on the record that this ship was launched something like a month or so earlier than originally planned. That was the lead [first, and hence most important] ship in a class of fifty ships—it was a big program. And the Iron Works agreed it could be done but, in so doing, they had people sand-blasting the hull at the same time people were putting the slipcoat on, and they contaminated the grease.

If you rubbed your hands along this slipcoat, it would take the skin right off. It was an awful substance. The grit penetrates down, it locks the grease, destroys the mechanism [that makes it slippery].

I asked the VP of Production who was responsible so we could look more closely at what he was doing, and the VP refused to give me the guy's name. He said, we've taken care of it. I said, okay. So I wrote a memo. I said how fortunate it was that John Wayne was there to assist us in recovering a very serious situation. In the event that ship didn't launch, we would have to reblock it, roll out all the cradle, and start all over again. Regrease the thing. Because the longer it sits on the grease, the less likely it will launch. It *never* gets better.

So Dean checks the hutchets, the ropes, the cables, the shackles, the state of the slipcoat. He makes sure all the grease irons are out from under the sliders. Actually counts them, counts the grease toggles, the little steel tabs that serve as spacers between the raised ribbon and the sliders. He has the surveyors take last-minute sights along both sides of the ship to be sure that any protruding sections of the four stories of staging that surround it have been removed according to his instructions. This and a million other things he checks and checks again. Because he is the Launch Master, the Maximum Leader of the launch crews, one of only four people who can postpone or cancel a launch if he's unhappy with the cra-dle, the weather, whatever, no questions asked.

It's 3:30 A.M. now, and a klaxon sounds. To Dean and Tom and Chris it marks the beginning of the night's real work; from here on in they'll be going like gangbusters until just before launch. To outsiders, guests here out of curiosity perhaps, the klaxon sets the heart beating a little faster, announcing the start of something they've never witnessed but heard much about, a tradition unchanged throughout the whole of Bath's shipbuilding history.

It is called a *wedge-driving*. For the year that the ship is taking shape on the ways, it does not rest on the launch cradle, which isn't even finished until a few weeks before launch. Instead, the ship's weight is borne by hundreds of stubby inner shores under it, by longer shores that angle in against the entire length of outer hull on either side, and by the centerline blocks running stem to stern underneath the lowest point of the keel. After the cradle is built, just before launch, the weight of the ship must be transferred gradually from these shore supports to rest on the cradle itself. The wedge-driving is the first step in that long process.

This is what these night visitors see and hear, perhaps what they feel, as they come to drive the wedges.

At the gate the guard waves them in; they collect white hardhats and black plastic safety glasses with wire mesh flaps on the sides. Better put on your birth control, the mechanics warn, meaning: *Those glasses are so ugly no woman would ever get near you!* The walk to the ways between the silent buildings is long; a single truck slows and passes in front of Dean's trailers. Overhead, big as a grand piano, the block and tackle of Crane #11 dangles, and the visitors hesitate, their heads going back as their eyes slide up the great white arm. Ahead, the geometric towers of staging loom. Two stories up, the lights begin to blaze. Heavy boots clang out on the iron staircase as they climb, then turn a corner, making their way through a covered chute that abruptly spits them out on the ways, where the cold river air makes itself felt.

All is black, the sky blotted out. Ahead, a flat wall of welded plates rises up, at first almost vertically, then in a steadily flattening curve, flaring out overhead to the very edge of the surrounding staging. It curves up and out, but also away, a vast, shadowy expanse of steel stretching nearly the length of two football fields down the incline of the ways, to the water. The eye searches for a landmark, and the feeling is of sudden disorientation, the landscape shifting all around, the world out of focus. Finally the image resolves itself, the realization dawning that the flat, twisting plane bending up and away above is one side of the bow. Magically balanced on the thin edge of its keel, the ship rears slightly upward at an angle, racing away down into the dark as if hungry for the river.

Details begin to register. The lower corner of the bow is missing, with a huge wood and steel structure in its place. The steel of the hull is cold and clammy to the touch. Each welded seam is different, some flush, even, regular whorls of once-molten metal joining the edges of the plates, others sunken or bulging, the bead rising and falling to its own rhythm. There are strings of numbers, a diagram, white-painted directions not yet covered over by the thirty thousand gallons of paint to come, battleship-gray for the hull and a deep maroon the color of old claret below the waterline.

The eyes travel down the ways towards the water at the whole length of hull, and the flat sheet reveals more gentle curves as the sharp flare of the cutting-blade bow, the sharpest of Vs, flattens gracefully into a gentle U until it reaches the stern, where the sudden, almost obscene eruption of the propeller shaft disturbs this seamless skin. The prop on this side of the hull is as tall as one of the Cape Cod houses that line the roads of so many Maine seaside towns, with five evilly curving blades like polished scimitars thrusting out from the hub, each shaft driven by a pair of gas turbine engines, the kind usually glimpsed hanging off the wing of an airplane outside the window of tourist class.

The whole ship is slightly raised off the building ways, leaning down towards the water at a barely discernible angle, its belly propped up on 180 separate stacks of wooden keel blocks from bow to stern. There are so many people around the ship, it's hard to move. Men in grimy boilersuits sealed at the ankles with duct tape move along each side and underneath, dragging lengths of hydraulic hose, the gas and oxygen leads of a plasma steel cutter, lights, air guns, sledgehammers.

The ship doesn't topple over sideways—though it appears to be about to at every moment—because it is held in a close embrace by a forest of angled side shores along its length, and supported underneath by stubby inner shores butted up against clips welded to the hull. Rising up from the bed of the ways at intervals almost down to the water, these huge supports of steel girders and foot-square wooden building shores have held her rigid and unmoving for the last year. Each is different in size and shape according to where it meets the hull, but all arc up gracefully, terminating in blocks of spruce and hemlock carefully molded to hug the curve of the ship. Tonight they will be disas-

Port and starboard, thirty groups, in teams of three to a ram, line up to drive the white- and red-painted wedges, snugging the cradle up to the ship hours before launch.

sembled as the launch cradle takes on the almost thirteen million pounds of the nearly completed ship.

Along each side of the hull from below the saddle down to the stern there are thirty groups of three men, each group resting a ten-foot, 150-pound iron ram on a rail about five feet off the ground and parallel to the sliders. Protruding between sliders and canting blocks are the wedges. Most of these wedge-drivers are volunteers from around the company; some are guests. Tom will supervise on the starboard side, Chris on the port, where a group drawn from the ship's first Navy crew will drive the wedges. Dean and Erik huddle all the while, observing, directing, Dean keeping an ear on the unrelenting radio traffic.

One team—David Bosse, Dan Athearn, and Steve Richard—are all pipefitters, all in their fifties, all hefty and solidly muscled from wrestling every day with the thirty miles of pipe that go into one of these ships. They're in a good mood; all around them fly crude jokes about poles and rams, guys smoke cigarettes, eat potato chips from a bag.

It's pretty special to be here, Bosse is saying. They've done it this way for a hundred years—it gives you a sense of completion. He shrugs, waves a gloved hand at the largely male groups waiting for the klaxon to signal the start of the first round of work. Most of these guys, they're all volunteers, he points out. It's something you want to do, makes you feel part of the ship. And anyway, we're not gonna have the chance to do this pretty soon anymore. All this—he nods at the ways, then at the line of men up and down the hull—it will be gone.

Down the ways Don "Lippy" Lippoth is watching his crew clown around. Shaggy black hair peeks out from under his hard-hat over a boyish face that smiles easily as his friends rib him.

Most of the time, we look out for each other, he says. You got heavy lifts going over you all day long. Every toxin known to man you're exposed to here. But I'm not one to fly a desk. I like the ships. They're an amazing piece of equipment. I believe in the Navy. Chances they have to go into harm's way are pretty slim, but if they do, I'd like them to have a good ship. Bath built is best built, we say here. You want to see that ship coming back up the river after sea trials flying all the brooms from the mast.

This last is said with a broad smile as he looks out at the river, remembering. This is an old custom in shipbuilding, flying new brooms from the mast as a recently completed ship steams back into port from her shakeout voyage. It signifies that the ship has passed all its tests with flying colors—a "clean sweep." Company directors especially like to see this, for it means the ship has per-formed to spec—and that the company will now get paid.

The klaxon sounds, and the men surge as one towards the hull, shouldering their rams. All up and down the seaweed-strewn ways, men grasp the bar as best they can, crowded in among their crewmates. Under the sea of moving hardhats, faces yet to see a razor and faces grizzled and red with effort harden in concentra-tion as each group rocks one step back in unison, then drives the head of the ram against one of the half-dozen red- or white-painted four-inch by four-inch wedges poking out in a horizontal line at the level of the rail. The oak wedge surges forward an inch at a time; then, when progress stops, the men take a delicate half step sideways at the same time they rear back, lining up the ram for the next wedge, then bang away again, moving up the line. The klaxon blares, and the first rally comes to an end.

From up close, the illusion is that these few men are lifting the ship, the incline of each wedge's face almost imperceptibly but inexorably forcing the hull up. It's not true, however romantic an image. In reality, the majority of the ship's weight is still borne by the blocks beneath its keel and by the angled side shores. Together keel blocks, inner shores, and side shores make a separate support system that has held the weight off the cradle until this moment.

The first step in weight transfer is tightening the tie-rod and wire assemblies holding port and starboard sides of the cradle together. Driving the wedges is the second step, transferring a little more of the ship's weight to the wooden cradle. Each wedge pushes the cradle tight up against the hull at the same time it drives the slider down into the layer of grease and hardened wax over the ways. So driving the wedges does not lift the ship, but instead snugs each section of support tightly up against the hull.

No one, however, can have any illusions about the noise. Talk is impossible as, up and down both sides of the hull, one hundred and eighty men slang away in the middle of the night, a kind of Futurist ballet of oddly dressed dancers. With sixty 150-pound rams crashing in a staccato rhythm and the squealing of the hydraulic wrenches, the steel hull booms and rings from bow to stern, vibrating, singing, the noise rolling off the steel in waves to lose itself over the water. Three minutes on, five minutes off when the klaxon sounds again, eight rounds from three-thirty to four-thirty in the morning.

Before the age of steam, a wedge-driving was a sight a visitor to any one of the dozens of Bath shipyards up and down the Kennebec might see, not twice a year, but twice a month when there wasn't ice on the river or if the weather wasn't too terrible. At the time of the Civil War, Bath was one of the ten most active ports in the country, launching more than two hundred and fifty ships between 1860 and 1869. It was a place of such strategic importance that the Union hastily constructed the defensive position at Fort Popham overlooking the mouth of the Kennebec River ten miles downstream to ward off possible Confederate invaders.

Today BIW is the last shipyard in New England (and one of only two in the entire country) to build ships of such great size and weight on the ways. With the launch of the *Donald Cook* in the morning light, it will be just a few hulls away from the end of this tradition.

Erik appears at Dean's side in hardhat and work clothes, says hello. Together they wander up and down, Erik poking his head in to see the wedge-drivers start the driving-up. Chris, with a crew of eight men up forward, is getting ready to begin the real work—taking down all of the keel blocks at the intervals and in the order dictated by the schedule. Erik's face is serious but relaxed, and Chris can tell that the naval architect likes the way the wedges are going in, how hard the men are hitting them, how much movement they're getting when they hit them.

Dean is happy, too, because the hundred and eighty wedge-drivers banging away *have* to put their backs into it in order for the next act to go on. He explains: You don't actually drive the ship up, you tighten the cradle up so it allows you to take your keel blocks and your ship support out. You try to transfer some weight, that's all you do.

Chris and Tom, who are waiting for the order to begin removing support, both know a good wedge-driving means the keel blocks come out easily. After all, the ship has been settling on the blocks, directly underneath the centerline at the hull's lowest point, for months, some for nearly a year, compressing them, making their edges disappear as they fuse into a single entity under local pressures up to a couple of tons per square foot.

The area that gets special attention during the wedge-driving is the trigger. There are two of them inset into specially constructed steel pits set several feet into the face of the port and starboard ground ways about ninety feet up from the shoreline. Each has a steel face that protrudes above the level of the ways, butting against a cutaway section of sliders. When the trigger arm is thrown, the butt rotates forward, below the level of the ways, freeing that section of sliders, and thus the ship, to move aft. The strain on each trigger approaching launch climbs from nothing to about 170 tons per side in a matter of hours, hence the care not to jar or otherwise disturb it.

BIW has never had any kind of trigger failure, and Dean wants to keep it that way. Which is why, he says, I make all kinds of checks on the triggers. Launch day, I don't have them drive up around the triggers hardly at all, just snug up the wedges. I don't want the pounding and stuff around them.

After two rallies on the wedges, at 4 A.M., Dean and Erik disappear into the gloom at the back of the yard, on their way to the Administration Building. In a nondescript conference room overlooking the parking lot and a cheap, greasy restaurant the mechanics call "Burger King," the two find Jay Bailey, the VP of Production, and Rick Libby, the Ways Director.

The last member of the team is still at home, in bed, having paid close attention to the weather maps on two of the three local TV stations and tuned his nightstand VHF radio to the National Weather Service report for Casco Bay and coastal regions before retiring. He is Captain Earl Walker, a former merchant mariner turned port pilot, the man who will be, by U.S. Coast Guard regulations, responsible for the ship from the moment it touches the water to when it either reaches open ocean or is safely docked.

The four gather around the speakerphone and talk wind, tide, and weather. This is called the "Go/No-go Meeting," and usually takes all of five minutes. The weather at launch should present no problem: the National Weather Service forecast is for 40- to 45-degree air temperature, winds out of the southwest at 5 to 10 knots, with overcast skies perhaps dropping a light rain or sleet later in the morning.

Time and tide wait for no man, the saying goes, and at BIW they wait for no ship. The bottom line is that the *Donald Cook*, given its launch weight of about 5,700 long tons, its five-hundred-foot length, and the shape of its hull, will need at least twenty-nine feet of water in the river—first, so its stern doesn't bottom out before it becomes buoyant, and second, so that the bow doesn't bang down onto the remnants of the cofferdam used to build the ways that lies submerged about eight feet just off the way-ends.

Why not launch at peak high tide, with the greatest volume of water moving upriver from Casco Bay giving the most leeway? In two words, the Carlton Bridge. This would be very dangerous.

The empty ship rides high out of the water, is moving in the river at one to two feet per second at its slowest, and has as yet no power of its own. Even a negligible wind out of the southwest (winds upriver are commonly 15 to 20 knots) combined with a modest tidal current of a fraction of a knot and an empty ship presenting a huge area of flat, resistant surface—this is a recipe for disaster, as BIW has discovered in the past when wind alone has nearly pushed ships into the bridge.

The net effect of these considerations is that launch time is invariably one hour after peak high tide, called *slack high tide,* when the great volume of water pushed up the river from the ocean is relatively still and unmoving. Sixty minutes after slack, the rush downriver begins, so even a delayed launch would eliminate or greatly lessen the possibility of hitting the bridge.

Tonight the men all agree that there are no significant problems with wind, weather, or ship. If they had decided to postpone the launch—which BIW has done exactly once in the past, in the teeth of a nor'easter—there still would have been significant shore support under the ship to take enough weight off the grease so the slipcoat would be preserved.

Down on the ways the wedge-drivers labor through two more rallies, as Dean calls up his supervisors. It's 4:15 A.M., less than seven hours to launch.

Chris, copy? Tom, copy? They respond in turn. *Chris, forward crew, take down bow to Frame 58. Tom, stern crew, take down stern to Frame 378.* For units of measurement on a ship, the Navy specs use a one-foot framework. So Frame 1 is one foot from the bow. Frame 505 is the very stern of the ship on the DDG-51s, which are five hundred and five feet long. The dismantling of the shore supports has begun, and from here on in, there is no going back.

The ways now become an even more dangerous place as both crews begin to knock out the shores and the blocks under the keel in the order dictated by the launch schedule, working towards each other in a precise pattern. Each crew has mechanics on the building ways, felling the tall spruce shores on the outside of the hull. Then the crews working underneath knock out the stubby inner shores. The ripping whine of hydraulic wrenches joins the wedge-drivers' banging as they begin on the keel blocks under the centerline, the lowest point of the keel.

Just as there is a precise order and function to the layering of blocks from slider to hull, so each piece of the centerline support has its function. On top of the stack and against the hull is the fitting block, cut to mimic the slope of the hull at any one point. Below this is a *quarterblock,* so called because it has been cut lengthwise corner to corner, creating four triangles that are then bolted back together to make a square. The height of the block is now adjustable, and it will sag naturally, while not collapsing, under the weight of the keel. Below the quarterblock is a *marrying wedge,* another foot-square timber with a single diagonal cut lengthwise, creating two halves that sit obliquely on each other.

The mechanics first remove nuts as big as a fist from the ends of the bolts that hold the quarterblocks together. With the wedges driven, usually enough weight is off the centerline so the blocks come out easily. If not, a few whacks with a sledgehammer against the marrying wedges are usually enough to topple them.

At the stern of the ship, meanwhile, a third crew has been preparing the end of the ways. With the tide out, the launching ways dip down, running out twenty feet before disappearing in the black water. Two men, one on each way-end, sweep flaming gas torches in an arc, back and forth across the surface, drying it completely before those behind apply a final coat of grease.

How they comin' out? How they dropping? Dean is asking Tom and Chris over the radio. How the first keel blocks come out gives him a good idea of what the crews will be up against farther down the hull. A good drive-up on the wedges transfers enough weight from blocks to cradle so that the blocks then come out easily.

If the ship's been there longer, Chris says, if it's heavier, they'll come out real hard. We've had the ship settle to the point where there's very little room for it to drop. We've actually dropped the quarterblock and there's still weight on it. All you can do is hit 'em with a maul, try to separate that marrying wedge.

When he hears the magic words—"They're comin' out easy"—Dean knows that eight rallies are enough, and will relay this to the man on the whistle.

The klaxon sounds two short, final blasts, and the wedge-drivers fall back, dirty, panting, and sweat-covered. Tom, his eyes forever traveling to the clipboard he carries like an amulet, gives the sign to move out. Each group of three shoulders a ram and, single file, they move up in a line along the hull, making their way wearily past the reviewing stand that will shelter the VIPs later in the day, disappearing into the dark of AB. Tom stays at the stern, checking that his crew is dropping the correct supports according to schedule.

Fifty mechanics still labor under the ship. They will work through the early morning until the ship is almost freestanding, alongside a crew of welders, gougers, and painters who move behind them, burning the steel clips that held the shores in place off the hull, grinding the steel down to bright metal, then painting it over with the dull brown of hull paint.

Since the day before, when the last of the grease irons were removed and the first tightening up of the cradle began, a mysterious process has been taking place, silently, almost invisibly, as the cradle embraces the ship, settling into its parts and in a real sense becoming one with it. It doesn't creak or groan, doesn't shift perceptibly, but the mechanics know. Tom, Dean, Chris, Erik—they know, too, though the only evidence is the thick ribbons of crayon-orange grease that begin to ooze from underneath the sliders.

They want the chance to let the ship settle down, Chris says. You don't want to dump all the weight on that launch cradle at once. You want to gradually ease it onto the cradle. When you drive the wedges, you lose a little bit of weight off the shores; when you tighten everything up, you lose a little weight. When you start taking your outside shores down, you lose a little more weight. Inside shores, a little more weight; centerline—same thing. You work bow and stern to midships, and you want to finish right about where the trigger is, because that's where all your weight is gonna end up.

There are fewer men around the ship now, and the quality of the noise has changed, too. There is the intermittent crash as two teams at either end drop the tall side shores, the whistle and chatter of the hydraulic wrenches more pronounced without the background of wedge-driving. Radios crackle and supervisors shout while mechanics curse as they wrestle the sections of keel block into neat piles out of the way of the cradle.

By 7 A.M. the last sliver of moon has disappeared. Clouds have begun to gather over the water, which has taken on a delicate shade of lilac from the lightening sky, the flat calm giving way to the slowly building rush of the incoming tide, the launch tide. The breeze freshens, smelling of rain, and the temperature begins to fall with the barometer. The ship waits on the ways, its millions of pounds settling onto the grease and wax with a steadily rising pressure, the weight on the triggers building from nothing at all to 50 tons, then 100, creeping inexorably upward.

From the rough sizing of the first steel plates to this moment, nearly two years have passed, two years in which the heads and hands of the seven thousand men and women of the yard have somehow taken a pile of cold steel and transformed it into something very much alive, a ship about to take to the water for the first time. Once she is off and gone, they will speak of her with respect and affection, her challenges and problems having become, for a time, their own, passing into their personal histories, stories to tell their grandchildren. How does this happen, that something so cold and so deadly—it is a warship, after all—and which they will almost certainly never see again, that this can take up such a space in their hearts?

LAUNCH

Of all the thousands and thousands of miles this ship's gonna sail in its lifetime, it's the first five hundred feet that are the hardest.

—Larry Albee, retired, Master on ninety-seven launches

If you want to get the best view of the launch from a place where you can truly see what is happening where it counts—at eye level with the cradle right beside the ship—the only place to be is next to one of the triggers less than a hundred feet up from the river's edge. Nearly the whole length of the ship passes before you, slowly at first, then very fast indeed. Dean Atkinson and Erik Hansen are there, along with some of the launch crew and the dignitary who will actually drop the trigger. Chris Medeiros and Tom Niles stand beside the opposite trigger, watching just as anxiously from their side.

After the last speaker's words have died in the wind, after the champagne has wet the bow, after the triggers bang down with a force of 300 tons behind them, there is a moment when everyone holds their collective breath, when the seconds crawl by. It's a lot like having a kid, Dean says with a sense of marvel in his voice, because, until you pull that trigger, you don't know what's going to happen. The number of things that can go wrong is nearly infinite, the launch crew's collective memories of things gone wrong in the past merely terrifying.

Dean Atkinson, Tom Niles, and Chris Medeiros out on the ways, launch day.

There have been what Erik Hansen terms "close calls." The December 1977 launch of the container ship *Maui* has gone down in BIW history as one of the weirdest, and certainly the most unsettling to watch on record.

Erik was at the trigger that day along with his then-boss, Ed Moll. The *Maui* was memorable even before its hair-raising debut. At 720 feet in length, 100 feet in width, and 10,000 tons in weight, it was a real monster, at the time the longest ship ever built north of Boston and launched in the Kennebec River. In planning that launch, the engineers' greatest concern was the moment the stern would begin to float, pushing 2,700 tons of weight onto a short section of hull just below the fore poppet. To distribute this great pressure so it wouldn't either buckle the hull or destroy the fore poppet, they used three saddles below the poppet, three bands of inch-thick steel each four feet wide, running ten feet up the sides of the hull from port to starboard and riding on blocking over the sliders.

Another thing they both remember is the bitter cold, a couple degrees below zero, which had set in weeks before and never moderated. The launch began normally enough. The trigger fell and the ship began to move, the stern plunging into the river. With the bow just past the trigger, the stern began to lift, and the moment the engineers had feared most arrived.

I heard a sound like a pistol shot, Erik remembers. My God, he thought, as huge chunks of concrete began to rain down on the ways. These saddles were made of wide bands of steel, with a stiffening layer of concrete poured into them. Under the sudden shift of weight forward, one of the saddles disintegrated, leaving pieces of concrete and steel fittings all over the place underneath a ship still not off the ways.

The forward saddle exploded, Ed recalls, a total failure. Just blew apart. With that, I could see the fore poppet teetering. I mean, when you lose one-third of your support! The ship went another fifty feet and the other two saddles exploded.

With no support, the fore poppet collapsed completely, just tumbled into pieces, almost at their feet. They watched, speechless, as the stem of the ship, completely unsupported, ground down the ways with a terrible noise and vibration. The force of buoyancy in the stern pushing upward and making the bow seesaw down onto the ways had compressed the solid part of the packing in the main section of the cradle as much as the wood could take. And then the great pressure had just blown that apart, too.

From there it got worse.

Ed remembers: These broken pieces of saddle are dragging up all of the keel blocks that have been removed from supporting the ship, and a *lot* of concrete. Now we're faced with the biggest drag that you've ever seen. Conceivably it could bring the ship to a halt. It will *never* get waterborne. If that ship didn't get to the water, it would just stop there—and capsize. Just roll over. It would be like the *Normandie* [which lay on its side next to the pier in New York's Hudson River after a fire]. We'd have a *Normandie* in Bath! I said a prayer at that point. *Please God,* I said, *let this ship reach the water.* And it did! The ship was doing one foot a second, creeping into the water.

As always when a mishap occurs, Erik says, it's not one thing, it's a combination of many small things. Indeed, the postmortem revealed that three different departments had each made one small mistake which, together, created this near disaster.

The steel in the saddles was of a poor quality, Erik continues, and the temperature the launch day was subzero. The saddles had been sitting out several weeks during the building of the cradle. Possibly an inadequate amount of attention was paid to getting the load spread out evenly between the three saddles. The crushing arrangement wasn't optimum, and because of that, the load was heavier on the forward saddle, which then failed, which caused the next saddle to fail, and then the third saddle to fail.

And that is how "mishaps," "close calls," and "near misses" (only the mechanics call them "fucking disasters") happen at Bath Iron Works, one foot at a time, from ways to water, with thousands of spectators and nobody able to do a damn thing to stop them.

While Dean Atkinson will fret until the *Donald Cook* is in the water, Erik, after thirty years with the company, is a little more sanguine. From his office in the mustard-colored Engineering Building, a converted warehouse that will never look like anything other than exactly that, he oversees, quietly and genially, a fiefdom of fifty naval architects and engineers. Officially he is the manager of Hull Engineering, with responsibility for all things to do with the structure of the ship, including all of the engineering arrangements involved in launchings, drydockings, and the onerous paperwork required by the Navy to document each hull's naval architecture, its attributes and behaviors, its stability.

If a man's office can be said to express his inner attributes, then Erik's personality must be conflicted. The plain, gray-brown steel desk is awash in papers, as is the flat-topped wooden cabinet whose shallow drawers hold hundreds of blueprints and drawings. The shelves behind his chair bulge with much-thumbed tomes like *Classical Mechanics, Numerical Control,* and *Principles of Naval Architecture*. The effect is of far too much in a too-small space.

The walls, however, betray a simpler, more reflective man, one who loves ships and the sea. There are dramatic photographs of various warships at launch and cruising, but also one of J. P.

Erik Hansen, chief of Hull Engineering, worries and wanders up and down the ways as launch time approaches.

Morgan's lovely yacht, the *Corsair,* and of the *Snorri,* a Viking ship meticulously crafted just down the river in Phippsburg for a recent re-creation of Leif Erikson's voyage to the New World.

One entire wall is an oasis for the eye, a photographic mural of a sheltered stretch of water and shoreline. Low pines grow along the banks, and where there is no green there is the gray of the rocky shore, or the slate of calm water. Much more than a holiday snap, the mural has something Zen-like to it, the emptiness and spare composition an invitation to the kind of deep reflection mirrored in Erik's calm demeanor, his quiet presence. He jokes with visitors that this idyllic place is where he moors his sailboat, down on the nearby New Meadows River. It is actually a photo of a pond somewhere in Oregon left by a former occupant of the office.

Launching a ship is not like horseshoes or hand grenades. There is no "almost." Either there is what Dean calls "a good launch," which means the ship goes off smoothly and on time, there are no leaks when she's in the water, and no one's hurt. Or else it's a total unmitigated disaster, which, thank God, has never happened at Bath. Were a billion-dollar ship to come to grief at launch, it would kill the shipyard. No bank would lend the tens of millions of dollars in floating credit BIW needs each year, no insurance company would bond them, the Navy would think twice about entrusting the construction and launch of even a lifeboat to their care. Quite simply, BIW could build the Rolls-Royce of ships, but no one would order it if the shipbuilder did not have a sterling reputation in the launching of ships as well.

It seems so simple. The trigger is released, and the ship starts to slide, stern first, very slowly, then more quickly down the ways. The stern splashes into the water, the ship floats free, the tugs move in to catch her up. That is what the launch spectator sees, all within the space of ninety seconds.

To understand the physical challenges of launching a ship at the yard, imagine a car parked on a hill with the emergency brake on and the engine off. Now release the brake. The car begins to roll, gathering speed. The challenge is to stop the car by means of a tow truck following behind it, trying to attach a chain. High winds buffet the car, pushing it from side to side, and a thousand feet ahead a train is approaching.

As a naval architect, Erik perceives the launch differently, in discrete moments, each of which presents its own challenges and dangers. The ship at rest, supported by the cradle that sits atop the inclined ways, is said to be at the initial *declivity*. As it begins to move, the stern, which hits the water first, slowly gains buoyancy and focuses more and more weight onto the length of hull and cradle still on the ways. At last, the buoyancy counteracts the force of gravity, lifting the stern, and the ship is *at pivot*. The ship then runs out of ways at *bow drop-off,* and the stem of the ship, its foremost part, actually plunges into the water. Finally, a tugboat takes up the slack on its towline and begins to swing the bow around so that the ship is perpendicular to the bridge, assisted by three other tugs, which are busy slinging various lines to help push it in for docking at the pier between bridge and A-Ways.

The first concern, Erik says in precise English slightly muddied by a Danish accent, is that when the ship is let go, it in fact will start—slide down. That's how the width of the sliding ways [the sliders] is determined. On the *Donald Cook,* we have four-foot-wide sliding ways. We had up to seven-foot-wide on the big commercial ships, and smaller ships, we had about two feet. It comes to having the pressure on the grease within certain limits. If it's too high, it will squeeze the grease out before you can launch, and it won't slide. If it's too light, for some reason the grease is such that it won't slide either.

We see the ways go out into the water, he continues, but what many people don't realize is that shortly after they are in the water, they also end abruptly. There's no more ways. And that gives a naval architect some concern—what happens when it runs out of ways? If the ship is not gaining buoyancy in the stern fast enough, it would rest hard on the ways.

It's as if the ship were sliding off a cliff. As more and more of the ship hangs out into space with nothing to support it, more weight rests on the very edge of the cliff, in this case the way-ends. Erik calculates the maximum way-end pressure—about 7 tons per square foot on the DDG-51s like the *Donald Cook*—to ensure that it won't damage ship, cradle, or ways. Too much pressure can also create friction, slowing the ship down so it doesn't get off, but stops on the ways.

As it happens, the wide and deep stern of the DDG-51s displaces enough water to provide good buoyancy, reducing way-end pressure. For other ships—the more needle-shaped FFG guided missile frigates, for example—enormous steel buoyancy tanks were attached on either side of their long and narrow sterns to provide additional buoyancy and also because of the limited strength of the hull over the way ends.

As the ship slides off the cliff, there will come a moment when the weight hanging over the edge is greater than the weight still on land. At that point, the ship will tilt.

Of course, Erik says with a dry smile, that would be very devastating. The center of gravity would be such that, if the bouyancy isn't picking up fast enough, it would physically tip the ship over the end of the ways. The next thing that would happen more likely than not would be the bow slamming back down on the ways after it gained buoyancy. We don't want that to happen!

Falling off the way-ends is not the same as pivot, however, for pivot is defined as the point where the buoyancy actually lifts the stern. As the stern lifts, about 20 percent of the total launch weight comes to bear, not equally along the entire stretch of hull still on the ways, but concentrated on the very front section of bow. Specifically, for a few seconds, the fore poppet under the bow must sustain 1,000 tons of pressure. This is the function of the fore poppet's crushing arrangement, the layers of plywood and white pine, to accordion at that moment, absorbing and dispersing the load as the plywood collapses and the pine compresses.

With the ship supported only by the fore poppet up front and limited buoyancy at the stern, strange things can happen. The ship will act as if it is grounded, and grounded ships are notoriously unstable.

To understand why this is so, first remember the basic forces acting on any ship in the water. Every object has a center of gravity, the point at which the object's weight can be said to be concentrated by the earth's gravitational pull. The center of gravity of a ship never changes, no matter where it is on the surface of the earth. Ships are designed with a low enough center of gravity so that the keel stays pointed down and the top of the ship remains upright.

Buoyancy is another force acting on a hull: a ship's buoyancy is equal to the weight of the water it displaces. The force of buoyancy can be thought of, in a sense, as trying to push the hull out of the water, a force anyone who has ever tried to submerge a piece of wood of any size has felt. Another way of thinking about this is that all of the air trapped inside the ship is lighter than the water. Begin filling the ship with water, which displaces the air, and at some point it will sink.

Things do happen very rarely to capsize a ship. For example, in a storm, upper compartments of the ship could fill with water, making the ship top-heavy. Were a leak to develop lower down, this, too, would decrease the ship's ability to recover because it would decrease the amount of air trapped inside the ship and thus its bouyancy. Sailing yachts lose their lead keels in violent seas. In all these situations, the center of gravity moves higher in the ship and can lead to the ultimate catastrophe, the boat capsizing, lying sideways in the water. (See Chapter 2 for a more complete discussion of center of buoyancy/center of gravity.)

On the DDG-51 destroyers the center of gravity at launch is higher than it would be on a fully fueled, completed ship, and thus the danger of capsizing during launch is higher. At pivot only part of the ship, the stern, is in the water. This presents only a limited surface area of hull against which buoyancy can act. Thus, were the ship to heel to one side, only a limited buoyant force would be counteracting the tilt. The ship could in theory simply flop over, heeling to such a degree that it could not recover and right itself.

On the *Donald Cook* and the other ships of its class, this danger is particularly pronounced. Due to whatever cause—a blast of wind from the wrong direction at the wrong moment, a panicked crowd of guests riding the ship rushing to one side of the deck, part of the cradle giving way more on one side and tilting the ship—the ship could begin to list and not recover. To mitigate against this happening, Erik directs that fuel tanks in the ship's bottom be filled with 400 to 500 tons of seawater before launch, lowering the center of gravity. Though the whole launch passes in little more than a minute and the pivot takes only five to ten seconds, ships have been known to capsize at this stage because they were inadequately ballasted.

Another particularity of the *Donald Cook*, Erik explains, is that when the fore poppet reaches the end of the ways, the ship is not fully buoyant, and the fore poppet therefore drops off the ways like a rock off a cliff. This could stave in the bow.

Because the ship is not fully floating, there is a considerable load on the fore poppet and thus on the way-ends beneath it. Erik doesn't want to see damage to either one. And when Erik says "drop," he means it. The entire bow plunges as much as six feet below the surface of the water as it falls off the way-ends, then surges back up. This could happen so quickly and so close to the way-ends that the very peak of the bow could smack the way-ends on the way down. With the ship moving even relatively slowly, however, the bow is far enough off the way-ends at the bottom of its surge.

The speed of the ship at launch is thus another worry.

Generally the ship accelerates from zero to between 15 and 18 feet per second in a matter of seconds, then slows just as quickly as the hull hits the resistance of the water. What Erik needs is a fine balance.

We're talking about 6,000 tons of weight drifting around out in the river, he says. We don't want it to hit the other side, and we do not want the tugboat pulled under. The tug grabs hold too soon, while the ship is still going at a good clip, they can't control it. They will be towed away. On the other hand, we don't want it to be so slow that it doesn't get off the ways, either. We need some energy [so that the ship will slide off the cradle]. By the time the ship is all dropped off and has gone maybe another fifty feet, the cable connecting cradle to ways tightens and pulls the whole cradle off, and we want enough energy in the ship to do that.

But the ship can't be going too fast, either. More than one to two feet per second, and the tugs begin to have real problems. With earlier commercial ships—which weighed twice as much as the DDG-51s and hence tended to go off at even higher speeds, there was a real concern that, by the time the ship had slowed enough for the tug to take up the slack on the towline, the ship would already be in the mud across the river or, worse, heading for the bridge.

The Carlton Bridge, whose on-ramp forms the northern border of the shipyard, looms large in the memory of anyone involved in launching ships at BIW. The Maine and Inland Coastal Railroad uses the lower deck just thirty feet off the water, while the upper roadway carries Route 1 the half mile across the Kennebec River from Bath to Woolwich and points north. Like the Green Monster at Fenway, it is ever present and immovable, an implacable thing of cold stone and steel that can turn an easy victory into a stunning debacle given just a few knots of wind from the wrong direction at the wrong moment.

We've had some near misses with the Carlton Bridge, Ed Moll recalls, especially with the States Steamship ships [four very heavy, very wide, and very long RoRo (Roll-on Roll-off) cargo vessels, which BIW built from 1974 to 1977]. One came within a hundred yards of the bridge. That was the *Maine* [a RoRo launched in May 1975]. We had barn doors on those ships to slow them so that they would be at about one foot per second at mid-channel. [Barn doors, technically called *masks*, are enormous steel plates welded perpendicularly to the stern to increase the drag of the ship through the water and thus slow it more quickly.] You can't get tugs to hook up to moving ships because there is a danger, if they hook up, they get tripped—pulled over, capsized.

In the case of the *Maine*, we tried to save some money. We used barn doors from an earlier launch, adapted, and told the port captain they would give him two foot a second instead of one. He judged that was okay. The result was that the *Maine* reached mid-channel doing two foot a second—and just kept going! They could not slew it.

Normally, as the ship heads out into the river, a single tugboat with a line already attached to the ship's stern takes up the slack and begins to haul the stern downriver, pointing the bow towards the bridge. Some of the speed of launch is absorbed in simply changing its direction—*slewing* it around.

Ed continues: The draft was so great, there was so much drag to that ship with all of its weight—the big deckhouse, the heavy machinery, the stern ramp—that it was just like a keel, just kept going and going—all the way to Arrowsic and Woolwich across the river.

As the engineers watched from beside the trigger, they knew something was wrong. The ship just was not slowing down, and the wind had stiffened, blowing upriver, towards the bridge. And why were the tugs still sitting there, as if stuck in the water?

Two tugs generally work the stern line attached to the ship. The first, the *Verona*, is a 2500 hp "power" tug that takes up the stern line and then waits out in the river in line with the ship's stern. Holding the *Verona* in that position against the current is the much smaller *Kennebec*. Just before launch, the *Kennebec*'s cooling intake clogged and her diesels quit. The *Verona*, pushed by the gentle but persistent current, slid slowly out of position, drifting upriver—towards the bridge. If the *Verona* had powered up after launch, she would have pulled the *Maine* even closer *towards* the bridge.

By the time the *Verona* had repositioned herself and powered up on the towline, the enormous RoRo was only one hundred yards from the bridge. One of the tugs powered around her bow and was prepared, in effect, to ram her, if necessary, to keep her off the bridge, and ended up *under* the bridge itself. The tug hit a rock, holing her side. By the time the RoRo was safely docked, the tug was sinking.

The *Maine* was the first of four ships in its class, and the yard did not want to repeat that particular thrill on the remaining three. Ed increased the size of the barn doors by 50 percent. They had meetings to discuss whether to alert the state authorities, to suggest closing the bridge. The yard finally decided against this because, if word got out, they'd have thousands and thousands of people come out to watch a BIW ship take out the Carlton Bridge. The launch went ahead with no further changes.

Pier 3 juts out into the Kennebec just downriver from the ways, and its north side faces them. The *Maine* had been moored there some months as her final fitting-out and sea trials took place, and there she sat, bow facing out, the next winter, when her sister ship, the *Arizona,* was ready for launch.

The first of November did not turn out to be a very auspicious day for a launch. Though the early morning forecasts weren't particularly alarming, by launch time it was a different story.

We had a very strong wind right out of the south, Ed Moll remembers. It funneled up the river. We had one guy with a home weather station, and he was watching it that day because he knew that we were in potential trouble. I think he clocked it somewhere between 30 and 35 miles an hour.

Now, the first ship, the *Maine,* which we had the problem with, was tied up at number 3 pier, north side. When the second ship went in the water, there was 15,000 tons of water displaced. You get this buildup of a wave, and then it progresses south. So the first ship is straining against the lines in this 30-knot wind. When the wave from the second launch got to the pier, it lifted the *Maine* up and rolled it slightly, and then of course as it passed, the hull dropped down and rolled out.

And every docking line snapped except for a lone hawser to the bow. The stern, following the receding wave, began to drift out into the river, ripping out every hose and line, including the banks of shore power and the welding gas lines. The sparking wires ignited the broken gas lines, and soon the base of the pier was ablaze. Thick clouds of brown, noxious smoke began to roll over the launch spectators.

As six tugs battled the stiff wind to keep the just-launched *Arizona* off the bridge, Ed watched the *Maine*. The stern crept around, then stopped, while a single nine-inch synthetic line keeping the ship from drifting away altogether stretched under the enormous load to the thickness of a strand of spaghetti. He waited for it to snap. The stern stuck in the mud, and the ship stayed there, straining against that single restraint, until a tug finally pushed up gently against its side, preventing it from breaking free entirely and drifting out into the river.

By 8:30, the sun is well and truly risen, and everything that can be done has been done. Erik knows from his instruments that all is well. The strain gauges on the triggers show almost 170 tons on each side, which is expected. Earlier, the engineers had attached creep gauges at the bow and above the triggers. These are no more than two blocks of wood with inches marked out on each, one attached to the cradle, the other to the ribbon, with their zeroes lined up. As launch approaches they will show that the cradle does, in fact, move down the ways almost half an inch as all the shore support is removed.

Everyone watches the gauges, and for a good reason. I want weight on the trigger, Dean says. And I do want some creep—it means the ship wants to go. If nothing's moved, that wouldn't be good.

Chris and Tom, the launch crew supervisors, watch the last shipfitter cut the electrical grounding clip from the hull, the final paint shop guy following behind to paint up the scar. The SUP-SHIP inspector is going over the bottom of the hull, making sure all the attachments were cut off and those spots made flush and painted.

From the bridge, Captain Walker alerts the tugs to stand by, then cast off. Six red- and white-painted tugs of various sizes, each with the white *W* of its owners, the Winslow family, on its funnel, begin to move into their positions, belching clouds of black diesel exhaust. The power tug comes in to pick up the towline, which

Crane #11 has hoisted and now slowly lowers to the tug's stern, where it is wound around the bitt.

On the ways, the men are pacing. Erik is pacing, Dean is pacing. Chris hears the speeches start and walks forward to listen with half an ear. He doesn't really hear them; it's just something to do to pass the time. Dean checks in with the launch crew on board the ship. One hundred lucky shipriders will line the decks as the *Cook* is launched, while Dean's crew of fifteen dashes through every compartment below the waterline with flashlights, looking for leaks. People always forget that, he says. The ship's never been in the water before. First thing you have to do is check for leaks!

An hour before launch, there's a good crowd already assembled, despite the lowering skies. Up front, the three hundred sailors who will be the first to crew the *Cook* sit stiffly in dress blues, white caps settled neatly on their laps. They trade smiles and nods with their parents, many of whom have come from far away for this important moment.

The *Donald Cook* is the first Arleigh Burke ship to be named, not for a naval hero like previous ships in its class, but for a Marine who fought in Vietnam, an unusual choice. As a result, various Vietnam veterans' groups have been invited, many welcomed by the yard as VIPs, with receptions and guided tours, and the stark black-and-white POW-MIA flag will forever wave from her mast.

At the back, the curious mix with the regulars. There are retired sailors, some of whom wear their brevets pinned to their raincoats. There are even a few tourists, looking dazed but happy to have stumbled on this unusual attraction on a rainy Saturday morning. Babies cry and parents shush older kids as latecomers stream in at the back, pushing forward in the hopes of finding a chair.

There is not much to see at this point. The spectators' chairs are laid out, not facing the ship, but at right angles to it, in orderly rows with AB on one side and the shipyard clutter lining the river-bank on the other. The ship's bow, draped in red, white, and blue bunting, juts up and out steeply over the speaker's platform. A security chain separates platform from audience, running from its sides left and right to close off all access to the ways.

The platform is an interesting construction itself, a white-painted raised dais with railings, chairs for about thirty under a wood-shingled roof whose center sprouts a small square stack vent like on the roof of a house. The effect is very New England, until a closer look reveals that, this being Bath Iron Works, the entire platform is of welded steel plate, the shingles a sham, the vent in fact a lifting eye. When the ceremony is over, the crane will pick the whole thing up and move it bodily back into the yard. The space it occupies, directly in front of A- and B-Ways, is far too valuable a piece of real estate to store it here.

BIW President Allan Cameron opens the ceremonies with a brief salutation. As he is speaking, a wheelchair motors into view on the roadway between the platform and AB. Its occupant has long gray hair restrained by a red bandanna and wears a fatigue jacket with the arms ripped off. He has no legs. The chair has been specially modified. A cupholder has been attached to one arm, and this holds a Bud Tallboy. The other arm sprouts an ashtray. He maneuvers the chair so he can see the speaker, his head swiveling from Cameron to the audience, a look of confusion on his face. He seems to listen a further moment or two, then continues on, across the front of the platform, to the empty space beside the sailors where the other chairbound vets have gathered.

As Cameron cedes the microphone to the next speaker, Dean radios Tom and Chris, directs them to do the final inspection. They hop down onto the ground ways under the ship, one on each side. They stop every twenty or thirty feet, eyes swiveling, making sure no one has fallen asleep under the cradle, or left a block up on the launching ways, checking that no wires or leads or hoses have been forgotten under the cradle. When the moving cradle encounters any forgotten object—which it has in the past—it obliterates the thing with a big noise, a noise no one wants to hear at launch.

Half an hour before launch Dean unlocks the padlock from the trigger handle and removes it. Erik waits beside him. The amplified speeches come to them in snatches on the breeze. Senator Patrick Leahy of Vermont is there, as is the entire Maine congressional delegation and Governor Angus King. King is ribbing the Vermont senator about the mildness of winters in his state compared with those in Maine, Leahy being the only man on the platform wearing an overcoat in the steadily deteriorating weather. It has begun to rain gently, and the wind is picking up considerably.

Out in the river the *Kennebec* throttles up a notch, its nose against the *Verona* amidships, holding the larger boat in a direct line off the stern of the *Donald Cook*. Broadside to the current, the *Verona* needs the smaller *Kennebec* just to keep it in position against the constant action of the tide, to be ready at any moment to move ahead with the ship, then head downriver, slewing the stern of the destroyer around.

Next to Dean, Jay Bailey, VP Production, is on the radio to the ship's bridge. *Are conditions favorable to launch at 10:44?*

Walker replies: *Affirmative.*

The seconds crawl by until five minutes before launch. *Speeches completed,* comes the message from platform to trigger crew.

Aye, Bailey acknowledges, *speeches completed. Request clearance to remove trigger safety pins.*

Walker is out on the narrow wing deck just off the ship's bridge, looking aft. He checks that the tugs are in position. *Request granted,* he tells the men at the trigger.

Dean moves into action, removing the secondary safety pins that prevent the trigger arm from falling before launch. He looks through a gap in the blocking to the other side, where Tom holds up an identical pin from his trigger.

Bailey: *Trigger pins now removed.*

Walker: *Acknowledge.*

Bailey: *Advising platform to proceed with invocation and that we are clear to launch when they are ready, confirm.*

Walker: *Acknowlege.*

Bailey radios the platform: *All clear—start invocation. When ready, signal to launch.*

The platform confirms: *Start invocation, aye.* Out of sight of the audience, a bottle of California sparkling wine is shaken mightily, then placed in a wire mesh sleeve that is anchored to a rope strung up to the bow.

Allan Cameron is the last speaker, a burly gray-haired Scot who has been building ships his entire life. He invokes the past, Phoenicians launching ships as BIW will today. He looks to the future, seeing in the coming modernization the shipyard's only hope for real salvation. The audience, more than a thousand people, doesn't appear to be listening. Their eyes are turned towards the great hulking flag-draped bow of the ship, whose very nose arches within feet of the speaker. The rain is heavier now, the wind more bitter. *Hurry up*, their faces seem to say, *and launch this thing.*

Mrs. Cook takes her place beside Cameron, who hands her the bottle and kisses her cheek. She faces the bow and holds the bottle aloft. I christen thee, she begins, her voice trembling, the USS *Donald Cook*. May God be with you.

The USS *Donald Cook* just after bow drop-off. The wax can be seen on the tops of the launching ways, and the bright buoys in front of the ship mark where the fore poppet is still floating.

She swings the bottle hard against the hull, and white foam erupts, the audience breaking into applause. Cameron hits a button on the podium, and a bell rings in the trigger pit. Donald Cook's eldest daughter, Karen Marie, with Rick Libby, Director of the Ways, at her side, takes the trigger handle. She pushes it forward, toward the river. Before it can complete even a quarter of its arc, the ship is moving, slamming the handle down with enough force to splinter the wooden catch pad. Rick hustles her out of the way.

The ship breaks free of the ways, and the yard whistle shrieks. The hundreds of spectator boats corralled downriver by a Coast Guard patrol blast their horns, too, joined by the tugs, whose stacks begin to belch thick coils of exhaust as they maneuver in. Cars stopped on the bridge sound their horns, and the launch crowd leaps to its feet, applauding, as the Bath Municipal Band breaks into *The Star Spangled Banner.*

Down in the trigger pits there is no cheering. Dean and Erik on one side, Chris and Tom on the other—these four men whose everyday existence is so intimately tied to these few moments—they know better. They wait and they watch, eyes narrowed, physically leaning forward, listening, holding their breath until the dangerous moments have slid by. Ship's going off awful fast, Dean is thinking as the cradle pops and cracks, minute imperfections in the launching ways magnifying the pressure in some spots, pulping the face of a slider. He sees the stern begin to lift and watches the fore poppet doing its thing, collapsing like a giant wooden accordian, the wood flattening, then coming back up a bit when the weight eases off.

Then the *Cook's* bow drops precipitously, and she is off and in the river. Now their eyes shift to the snubbing pits where the cables attached to the cradle are anchored, watching the slack cables tighten suddenly under their 100-ton loads. Fifty feet into the river, four fluorescent buoys shoot out of the water, marking the barely floating mass of the fore poppet. Just beyond, the bow of the ship roils as a few sliders shoot out of the water, then the whole cradle comes up in a rush on the upriver side of the bow.

Standing on the starboard wing deck of the moving ship, Captain Walker tracks the tugs, his eye straying aft to the *Verona,* whose stern boils as she moves with the *Cook* out into the river. If the *Verona* moves too slowly, the slack of the stern towline could wrap itself around the rudder. If she moves too fast, the line will snap. His eyes swivel forward, and he judges their speed by how far he is from the ways. Finally he radios the *Verona* to take up the slack in the line and move downriver.

A few seconds later he feels the deck shift slightly, the beginning of the slow turn that will swing the ship parallel to the riverbank, thus reducing the force of the current and the effect of the wind, both of which are hitting the ship broadside. As he watches their progress towards the bridge slow, then stop altogether, he sighs inwardly in relief. He wouldn't want to be known as the first person to put a ship onto the bridge.

Captain Walker calls in the four remaining tugs, who immediately begin making fast to the destroyer. The *Alice Winslow* butts her nose against the bow, the *Marjorie Winslow* against the stern, while the *Gordon Winslow* ties off amidships facing out into the river. At his order they power up, beginning the not uncomplicated maneuver of docking the *Cook* at Pier 1, just south of the bridge. Meanwhile the *Kennebec* has caught up the hundreds of feet of roped-together cradle and begins to tow it downriver to the storage yard for salvage.

The launch crew emerges from beside the triggers and gathers at the end of the ways. Dean does not cheer, nor Erik rejoice. Chris and Tom do not clap or shout congratulations. Instead, there is a circle of quiet smiles, nods of recognition, a clap on the shoulder.

While the dignitaries disappear for a champagne reception at the Atrium Hotel in Brunswick, Erik trudges off to collect instrument readings that have captured the ship's acceleration and speed at launch for the postmortem launch report. Dean, Chris, and Tom are all heading home, but not before making sure the second shift knows what's expected of them. One crew with shovels will begin to clean up the mess of grease and wax left behind on the ways and then bare the concrete with a high-pressure hose. Another will pull the salvaged cradle from the river off the storage area, beginning the long process of disassembling it and inspecting each piece for damage. The first keel units of the next ship, the *Decatur*, are already crowding AB, and the mechanics are going to need someplace to put them.

FROM HULL 463 TO USS *DONALD COOK*

When I joined the Navy, they wanted me to be a YN, a PN—a clerk filing papers, typing. I said no way, I want to work with weapons, with explosives, I want to sink ships, something you can't do on the outside, you know, working at McDonald's.

—*Torpedoman First Class Valarie Williams, USS* Donald Cook

In the year between the key milestone events of launch and sea trials, Hull 463 undergoes a transformation at first almost invisible, as the ship's interior machinery and electronics spaces are finished out, with thousands of miles of pipe and electric cable, fiber optics, and other wiring run the length and breadth of the ship and finally connected up to their ultimate terminus at a console, gauge, pump, faucet, switch, engine, motor, fan, switchboard, computer, telephone, toilet, control panel, LED indicator.

The first step in this transformation comes two months after launch, when the ship is towed to BIW's mammoth blue-painted, rust-streaked drydock in Portland Harbor. The ship enters the sunken drydock in the evening, and then, through the small hours of the morning (to take advantage of cheaper off-peak electrical demand) the water is pumped out of its hollow walls. By the next morning the whole structure, looking like nothing so much as an open-ended box with a ship stuck in it, rises completely out of the water, once again high and dry up on blocks.

The USS *Higgins,* DDG-76, gets new paint after the installation of its sonar bulb, the white-wrapped structure at the base of the bow.

Here the ship is given another coat of paint, but more important, the bulbous sonar dome that protrudes out and down from the base of the bow is attached. Because it projects so far below the rest of the keel, it would make the ship very difficult to launch on ways because the solid packing and saddles and poppets would have to be made that much taller, rendering the ship too impractical for launch. The sonar bulb is also an inherently fragile structure that might not survive the rigors of crashing down into the water at the point of bow drop-off at launch. The exterior is a delicate rubber sheath, which holds its shape because it is pressurized from the inside, creating the optimal conditions for the rows of spike-like sonar sensors that are aimed to detect underwater contacts everywhere but directly behind the ship.

The *Donald Cook* passes three contractual milestones during this time, three large paychecks for BIW, and three huge steps forward towards its maiden voyage under its own steam.

The first milestone, in January 1998, is ALO—Aegis Light-Off, when the approximately twenty-five component subsystems making up the Aegis combat system are brought on-line. Each and every part of this system—from the enormous SPY radar arrays at each corner of the main deckhouse to the smallest switch controlling a cooling fan down in the Combat Information Center—all must be rigorously tested individually and as they interact with each other. Three major computer systems—the Command and Decision computer, the Weapons Control System, and the Radar Control System—together with all of the software aiding communications and controlling the functioning of the actual weapons systems themselves, must be systematically put through their paces, too. Some idea of the complexity of these systems can be seen in the photo of the CIC in Chapter 1, simply by looking at the thicket of cables several feet high hanging from the ceiling above the controllers' consoles.

Three months later comes GELO, or Generator Light-Off, the second post-launch milestone, when the electrical power generation, conversion, and distribution system comes to life. Electrical power on the DDGs is generated by three Allison gas turbine generators, each of which produces 2500 kilowatts of electricity. Two of the turbines are located far belowdecks, one in each engine room, with fresh air intake and exhaust running in tortured twists down from the stacks above the deckhouses.

These systems are designed with some special considerations. Replacement of any generator or propulsion turbine, for instance, is accomplished by drawing the defective turbine up through its own exhaust stack on special tracks that line the stack's insides. To reduce the infrared signature of the ship generated by the heat of exhaust, the uptakes are made of exotic heat-dissipating materials, and the hot fumes mixed with large amounts of outside air. The result is that, even with the ship under way at significant speed, it is almost impossible to smell the ship's exhaust or to see it emerging from the stacks.

MELO, Main Engine Light-Off, the last milestone before sea trials, takes place usually within a month of GELO. To provide redundancy in case of a direct hit, the DDG-51s have two engine rooms, one forward and to port, one aft and to starboard, with each at a different level. Each engine room holds two General Electric LM–2500 marine gas turbines, similar in essence to the turbines that run a 727 jetliner. Both turbines connect to a single General Electric main reduction gear which, like a car's transmission, translates the very high revolutions of the turbine into the much slower revolutions required to turn the hollow steel propeller shafts. These run through a thrust bearing (which absorbs the thrust of the turning screws) to the seventeen-foot-tall propellers.

All parts of the propulsion system—fuel pumps, air intake and exhaust, turbines, reduction gears, shafting, propeller blades— must be tested. Noisy bearings or small misalignments in the shafts or props make for a more detectable, and potentially fatal, acoustic signature with the ship under way in hostile waters.

All the while the ship's insides are being put together, externally the deck spaces seem a rusty mess of tools, steel, and boxed parts. There are huge gaping holes in the deck fore and aft into which the Vertical Launch Systems will be set in place and covered over, and smaller, almost random holes in bulkheads and underfoot awaiting the overburdened fitter's or welder's attention. The painters appear, and at first slowly, then more and more rapidly, mauve primer gives way to haze gray. Seemingly overnight the USS *Donald Cook* emerges fully formed like a butterfly from its coccoon, a warship ready to steam away to ports unknown to take its place as the newest addition to the U.S. Navy's surface fighting fleet.

The end of Charlie Trial marks a final turning point in this evolution as hardhats give way to the navy-blue *Donald Cook* ball caps of the nascent crew, which is just beginning a period of brutal training and preparation that will give them the power to make this ship their own.

One part of this evolution is the noticeable, natural winding down of BIW's role as builder, the numbers of mechanics underfoot steadily diminishing as the captain and his chiefs gradually take over the ship compartment by compartment. A less obvious process has been occurring all along: the BIW workers and systems subcontractors, mostly electricians, technicians, and engineers, have been imparting to their Navy counterparts an invaluable trove of knowledge as the two groups work side by side to get their systems installed, tested, and up and running. This unique partnering is just the first step down the long road of melding crew and ship into a single highly functioning fighting unit.

During the Second World War, one of the lesser-known advantages the U.S. had over the Japanese navy throughout the long Pacific naval campaigns was not a type of weapon, or even necessarily superior ships. Rather, it was a particular skill—damage control. The ability of the U.S. Navy to contain volatile shipboard fires, then repair, refit, and return damaged ships to battle in a short period of time was a constant source of unpleasant surprise to the Japanese naval commanders.

Lieutenant Commander Mike Anderson, the *Donald Cook*'s combat systems officer, tells an anecdote about his assignment several years earlier as an Aegis liaison aboard the Japanese destroyer *Kirishima*, which is similar in design and combat systems to the DDG-51s. When we came by the island of Midway, he says, they had a very elaborate ceremony, a day of remembrance and training lessons—history lessons. They very much credit the Battle of Midway and to a large extent our winning the Pacific War with our damage control capabilities, ships that they thought were out of action showing right back up in battle. The U.S. Navy has had a strong emphasis on that, training being a huge part of it.

This tradition of superior damage control the Navy takes very seriously, and it begins with every individual sailor's ability to act as part of a team, as a crew, and ends with the superior knowledge all accumulate during their stay at BIW, the hands-on experience with their systems and equipment that allows them to fix what is broken when the ship may be under way weeks from any port. Like all navies, the U.S. Navy teaches shipboard firefighting and damage control through extensive training, simulations, and constant drills, which begin in the first days of boot camp and follow every officer and enlisted rating throughout his or her career.

Where the U.S. Navy differs from others is in the wholesale involvement of BIW and its associated weapons systems contractors (Lockheed Martin, General Electric, Raytheon, among oth-

ers) in the rapid and intense integration of the crew into the workings of the ship and its systems. Even before many of the *Donald
Cook*'s youngest sailors had even enlisted, before all of the steel
had been burned, before a single unit was erected on the ways, the
captain and some of his department heads arrived at BIW to begin
work. Nothing speaks more loudly to this than the simple fact of
geography: the *Donald Cook*'s Pre-Commissioning Unit—offices
and conference and training rooms for up to two hundred officers
and crew—are located in a three-story brick building just across
Washington Street from the yard.

Manning proceeds in an incremental, top-down fashion. A
year before launch, Lieutenant Commander Anderson says, we
had a dozen chiefs up here establishing a first interface with the
yard, getting to know everybody, and chiefs and department heads
getting together with the captain. You start laying out the command policy—what are the captain's philosophies, how are you
going to implement them? You've got to get the structure in place.
You can't just take a couple hundred sailors and drop them in the
middle of a small town in Maine and say, okay, go! Be a crew!

Anderson explains that the value of this on-site, hands-on
approach of introducing the crew, especially the technical staff, to
the ship as it's being built cannot be underestimated.

I had some exposure, he says, to the Japanese shipbuilding
process, a Japanese shipyard. The crew is not involved in the
build nearly as much as we are at BIW. They are kept off to one
side and trained in a schoolhouse. When they walk aboard the
ship for the first time, it's built, it's painted, it's labeled, the floors
are swept, the passageways are waxed. It's like a new car out of the
showroom. The Japanese really keep a strong hammer on the
shipbuilder in terms of level of detail and build, but the crew loses
a lot in terms of knowledge you can gain seeing the gear come up
on-line, going through the test process from the ground up.
Getting that one-on-one with your counterpart out of the shipyard who's going to be this incredibly knowledgeable individual
who's put this system together, from nuts and bolts and wires, a
half dozen times.

And, he continues, these BIW guys and the subcontractors
are terrific, outstanding about taking us under their wing. The

folks here at Bath have a lot of pride in their work, they know their jobs extraordinarily well, they know the equipment. They like teaching the sailors, showing off their stuff.

Our enlisted personnel, he continues, our technicians, come up staggered at various times, a third of the crew nine months prior to sailaway, half six months prior, the rest a couple of months prior. As soon as they get up here, the *first* place they go is with their chiefs right over across the street to the ships. See what phase their systems are in, see where their gear is. They can put their hands on it, touch it, see it. You know, the biggest thing you can do is just stay over there, stay on the ship. Find your BIW counterpart and start picking his brains. Let him teach you, and just watch it come together, Anderson finishes.

It is a huge challenge to forge that first group of approximately two hundred and fifty enlisted sailors into a functioning crew in six months, and to do this on a ship that is static and incomplete, tied up at the dock except for a grand total of perhaps seven working days at sea. It is like trying to train a racehorse locked inside the barn. The task is all the more difficult because, while the twenty officers and most senior enlisted will have been to sea, from a third to a full half of the crew may never have served aboard a ship even though some of them have been in the Navy for years. Much depends on the captain's abilities and his prior experience gleaned from years of service at every level up the chain of command to his present position.

By the time the *Donald Cook* sails away from BIW en route to Rhode Island to load ordnance, and then on to Puerto Rico for weapons testing and certification, it will have soaked up more than a billion taxpayer dollars. Tens of millions of dollars go into the last phases of testing and trials alone. By one rough deckplate estimate, any one sea trial—just getting the ship up and running, cast off, down the river to sea and back again—burns up a million dollars by the time tugboat and piloting fees, BIW man-hours, and food, fuel, and other overhead expenses are figured in. In the Navy, the DDGs are known as expensive real estate, and as a *new* expensive piece of real estate, especially one armed with so many offensive weapons, it requires what the ship's first captain drily calls "adult supervision."

Of the two to three thousand officers who emerge in any one year from the handful of schools which graduate officer candidates, usually fewer than one in twenty will ever be offered the command of a surface line ship, a naval combatant like this destroyer. About two-thirds will choose subs or aviation. Of those remaining, some will not have the right combination of training and experience, while others will have a disqualifying blemish or two accumulated over the eighteen to twenty years of service necessary to even be considered for surface command.

The ship's first captain is twenty-year veteran Commander James McCarthy, an intense, reflective man whose short, compact frame seems barely able to contain the energy and good humor that sparkle out of the warm blue eyes and animate the smiling, freckled Irish face. Once you get named CO [commanding officer] of a ship, he says, it's like you get energy from all kinds of places! You've achieved something. It took twenty years to get here, and lots of good guys fell off the table along the way.

On a brisk day in mid-September, with the crew just having moved aboard and yet sailaway only six weeks off, Commander McCarthy sits at the workstation squeezed into one corner of the private wardroom that makes up half of his palatial quarters, the other half a ten-foot by ten-foot bedroom mostly taken up by a duvet-covered bed that faces a wall-mounted TV/VCR. Privacy and space on a ship of this size are precious in any amount, and it is a measure of the captain's importance that he has, by right and necessity, the only truly private quarters on the ship.

Strategically located high up in the main deckhouse and so handy to the bridge, his space, though windowless and low-ceilinged, is nevertheless cheery and warm. The wardroom, perhaps twelve feet by fourteen feet, is a bright place, white flocked wallpaper and blue carpeting underfoot, with six green baize-covered chairs surrounding a table, white metal cabinets with locks, and the soft light of wall sconces. The workstation is cluttered with paper on every surface not taken up by both a Dell and a Hewlett-Packard computer. Above, classics like *Huckleberry Finn* and *Ivanhoe* in leather-bound editions share shelf space with *Gray Ghosts and Rebel Raiders*. One wall is dominated by a dra-

Captain McCarthy quizzing Signalman Chewning during Fast Cruise exercise a few weeks before sailaway.

matic presentation watercolor of the *Donald Cook* at sea by the well-known naval artist, Peter Hsu.

The captain leans back in his chair, hands clasped neatly behind his head, grinning out from under a mustache, which, like the profuse hair on his head, is of a fine heather flecked with gray. It takes, he is saying, a significant change in the level of dedication when you go from being the guy just out of college and doing his first—obligative—service to the guy who has made the decision, I'm going to sea *for the rest of my life.* I'm going to miss my wife's birthday and my kids' birthdays, and Christmases and New Years because I love going to sea that much and because being in command means that much to me.

Commander McCarthy leans forward, his eyes suddenly flashing, his voice intense, and the visitor shivers, getting the merest, searing glimpse of what it must be like to get called on the carpet in front of this man. The U.S. Navy, he intones, has put me in charge of a billion-dollar physical plant that's got three hundred people, moved me in my career sixteen times, paid for the education in several different instances to get me into this position. It's public record that they're paying me about $70,000 a year. If a Fortune 500 company invested that much in someone and was putting him in charge of a billion-dollar physical plant with three hundred employees, how much do you think they'd be paying *him*? So what I'm doing this for is *because* I get to sit in this chair. It's not because I'm getting paid enough. Part of it is patriotism, yes, I love my country. But the work doesn't get any better.

When I came on active duty, I had three choices: I could be a submariner, a surface fleet guy, or an aviator. I'm a surface nuke, by the way—a nuclear engineer for surface vessels. Everybody claims that submariners, they are the elite force, the best in the business, they can do anti-submarine warfare way better than all us surface fleet guys. I hope so, McCarthy laughs, because that's all they do! Okay, aviators—boy, that's the life for me, fly at Mach 2 with your hair on fire. But . . . you also don't have anybody working for you.

And working with people, for people, the captain continues, I think probably the most gratifying thing I ever had happen to me as a junior officer was when some eighteen-year-old kid who'd enlisted got out of the Navy [a few years later]. I got a phone call from his father saying how much he appreciated the fact that we'd taken his son and turned him from this kid who was going south to a man who could be relied on, who had integrity. That's what surface line duty brings to the table. It's not just driving a ship, it's working with three hundred guys out there.

The commander smiles suddenly. I own a sports car, he says, as if making a confession. And this is as close as it gets to having a sports car on the water. This is unbelievable. It's all degrees of happy here, the best technology, the best capability, lots of horsepower.

While the command of a new ship comes with a lot of responsibilities, it also comes with an enormous amount of freedom in how to meet them, how to organize things, with perhaps even more latitude than usual given to Arleigh Burke captains. The Navy recognizes the special nature of the Arleigh Burkes, a unique class of ships whose torpedos, Harpoon and surface-to-air missiles, and five-inch gun give it the ability to carry out anti-submarine warfare, anti-surface ship warfare, and anti-air warfare. Its Tomahawk cruise missiles lend the capability of striking targets far inland. All of this offensive capability does come at a cost, however, for each system requires physical space both for its equipment, communications, and apparatus; and for its operators.

Some senior surface fleet officers have been heard to grumble that this class of ship is a prime example of political decisions based on cost negatively affecting the size of a ship, something that should be dictated only by its mission requirements. Simply put, some wonder if, in the thick of a prolonged hot war in which many if not all of the ship's offensive and defensive weapons systems would be necessarily engaged, the smaller crews of these destroyers would, with little relief over days or weeks, become overtaxed and overtired.

Unlike the larger, more heavily manned ship classes, every member of the crew of the *Donald Cook* generally must perform more than one job, has so-called collateral duties. The physical space is tight, too. The enlisted sailor has a "rack"—a bed, two feet six inches wide by six feet six inches long—and racks come stacked three high in a compartment with a ceiling that is lower than ten feet. Elsewhere on the ship, two of even the thinnest of sailors cannot pass in most corridors without turning sideways, and between-decks ladders are so steeply pitched that you have to bend backwards while descending so as not to hit your head. All of this means that the crew has to get along, and get along well for periods as long as six months at sea.

Though Commander McCarthy did not handpick his crew, he did have the option to refuse a crew member assigned to him. This he did not do, preferring, as he said, to play the hand he'd been dealt, a hand with some unusual cards.

After the briefest conversation with any random sample of crew, the visitor comes away amazed at the disparity within the group; they come from all parts of the country and have widely differing educational and family backgrounds. Some speak with accents not only regional but foreign, the words of others move with the lilt of an urban rapper. African-American, Asian, Hispanic, and Caucasian, male and female, rural and urban, teenager and old salt—the poten-

Quartermaster Mate Second Class James Sippel on the port wing deck checks over his course indicator on a Fast Cruise exercise.

tial barriers to a smooth integration into a working crew seem many. Yet they are an excited, dedicated, talented bunch, most of whom seem to share the goal of their captain—to distinguish themselves as the first, best crew of the *Donald Cook* and thereby honor his name. This sentiment comes through very strongly at all levels, from deck seaman on up.

James David Sippel, at twenty-three, will have been in the Navy for one year come December 1998, when his ship sails away from its commissioning. He is an E-2, an ordinary enlisted seaman, a quartermaster. Quartermasters are, he tells you proudly, the ship's navigators and pilots, the captain's right-hand men, the ones he has to trust. We have the course we set in port laid down on the chart. When we are out in the ocean, the quartermasters' watch responsibility is to keep on track and let the bosun's mate of the watch and the officer of the deck know anything, any changes in the ship's course.

Sippel never finished high school. He was a skinhead, but an anti-racist skinhead. He was also an Eagle Scout, and got arrested at sixteen for spraying grafitti. He was married just out of boot camp and is now separated from his wife. He knows everything about the Titanic, its construction and sinking, and has his own theory of how the ship's captain (and quartermaster!) might have saved it. His nickname on the streets of Morristown, New Jersey, where he grew up, was "Rocky," because, he says, I'm a fighter. I've always fought for what I want, and I had always wanted to join the Navy. I didn't have anybody to help me get there. I had to work on it all on my own.

Sippel looks like a fighter, if perhaps a very young fighter not often needing a morning shave. He's short and scrappy, thick in the shoulders and short in the neck, on top a black buzz cut, with heavy black eyebrows, the left split by an old scar, over surprisingly soft brown eyes. Another pale scar mars the skin under his right eye, and his generous nose shows the crook of an old break. His arms are muscular under his everyday blues, one shoulder blazing with a bright green shamrock tattoo proclaiming "Irish Pride."

His story tumbles out slowly, simple words spoken with a quiet force belying the strong personality that got him where he is today. I was born to a hooker, he begins, and I was taken away from her when I was seven. She died of AIDS in '85. I was adopted at nine, and I later learned that my parents adopted me for the wrong reasons. My mother had a lot of issues in life, he observes with a compassion greater than his years, and my father wanted me to be a certain way and I wasn't. One day he just took me over to my grandparents and expected them to take care of me. Which they did—they are very, very loving and very giving people, very religious.

When I was smaller, the young man continues, I loved sailing ships, boats. I used to look at a picture of my grandfather in his dress whites—he was on a battleship in WWII. I used to be fascinated by it, and I thought the Navy would be an excellent choice of service because I didn't want to die, and a Navy ship doesn't go down without a fight! I like being a quartermaster because it's old-fashioned; the quartermaster has been on the ocean since the beginning of sailing.

I was living on the street, in a men's homeless shelter in Hackensack, New Jersey, when I joined the Navy. There was a recruiters' station right across the street. I had been in a lot of trouble with my family at the time, breaking up with a girl, this big [rift] between me and my grandparents.

Up on the bridge, Sippel says, that's my home. I work with the captain every day. Every day he comes up and says hello to us. He's the kind of captain you want on a ship. There's screamers, and then you can get a captain like James McCarthy who likes getting under way, going out there in the ocean, just sailing. He's a very stern and fair man, the judge and jury when you've done something terribly wrong with your life in the Navy. Sometimes he's served on the food line, asked me what I wanted and put the food on my tray. A very humble thing to do, Sippel says, a touch of awe in his voice.

On a warm day at the end of September you could find Sippel out on the port wing deck just off the bridge, Captain McCarthy at his side, the faces, attitudes, and body language of the two men all communicating the transmission of knowledge from teacher to student. The two stand next to the instrument that Sippel, as a bearing taker, will use in the course of navigating in any restricted waterway or harbor, a gunmetal-gray course indicator that helps determine the ship's location relative to fixed points on land.

The captain leans in close, directing Sippel to shoot three sights to take the ship's bearing by triangulation. Now I want you, he is saying, to shoot this flagpole, that bridge, and that lightpost. Which are you going to shoot first? The flagpole, sir, Sippel responds confidently, a pair of bulky gray headphones for communication with the bridge dangling around his neck. The captain nods for him to continue, and the younger man earnestly explains his reasoning. The captain listens, his eyes on Sippel's face, leaning slightly forward in a posture of complete attention. Sippel finishes and looks up expectantly, earning a curt nod and a smile that signals *well done,* the memory of which Sippel will live off for the next week.

With only weeks to sailaway, when the ship will leave BIW with its first Navy crew at the controls, every day is caught up in ever more intense training. This morning's exercise is called a

Fast Cruise, in which all departments simulate every order, action, and motion they will have to execute for the ship to leave the dock.

After completing pre-underway checklists, a four-hour exercise in itself, events unfold at a furious pace all over the ship. The watch—the small group of senior and junior personnel detailed to monitor the ship's systems and control access to it—shifts from fantail to bridge while the PA echoes with the phrase, *Manned and ready for sea!* as the heads of the various departments report in to the "XO," the ship's executive officer, or second in command.

The Special Sea and Anchor/Navigation Details are stationed at the fantail and amidships where the mooring lines will be cast off, and on the fo'c'sle as well, where a dozen deck seamen mass at the anchor capstan and console to walk through an exercise of making the anchor ready to let go. Bearing takers like Sippel appear on the bridge wing decks, as do lookouts, and suddenly the bridge is crowded with officers and crew gathered around the steering and throttle console, chart table, and radar screens as the simulation gathers speed.

Clipboards and checklists abound, as does the nervous tension arising from every sailor's knowledge that they'll be doing this soon, for real, with the turbines fired up and the props spinning, 8,000 tons of steel loose in the river with only a few spindly tugboats to assist if they screw up.

The captain makes the rounds, checking in with the helmsman, the navigator, those on watch on the bridge. He asks questions and answers them, offers encouragement and advice drawn from his twenty years of experience. Out the front windows of the bridge, the deck seamen are handling the thick mooring line, laying it out in long concentric loops to avoid snarling. The captain glances down at them; he'll appear unexpectedly all over the ship in the next hours, taking the pulse of his crew and officers, judging their incremental progress towards the coming test.

The ship is tied up port-side to the BIW quay, and a lines-handling team assembles on the fo'c'sle, port-side, for line hauling. Though they are working with state-of-the-art woven Kevlar line six inches in diameter, any line under tension is still dangerous stuff. When these lines snap with the weight of a ship behind them, not an unknown occurrence, the spring-back—the whip-sawing of the line back on itself—can take off fingers, hands, even arms. For this reason the crew of a half-dozen men and women wear blue hardhats and safety goggles, with their pants tucked into the tops of their boots and gloves on their hands.

First Lieutenant Friend, a sturdy blond-haired officer who brooks no nonsense from the boisterous deck seamen, wastes no time after the PA blares, *Take in all lines!* Take up line one! she orders. Don't let it drag. Keep the line off the deck. Watch the twist in the lines! she calls as the gray-white line comes in over the side to be coiled in long ovals on the deck.

Another team is simulating anchor drop at a predetermined location, with a lookout calling out distance to anchorage as the ship, in theory, approaches. Two thousand yards to anchor! the lookout calls. Two thousand yards, aye! the bosun's mate running the exercise responds. One thousand yards! Five hundred yards! How can you tell when you're stopped dead in the water? he asks suddenly. No wake! the seamen respond. Un-hunh! the mate affirms. And you're gonna be throwing over pieces of wood, bread. When you see that bread just sittin' there, you know you're dead.

The imaginary anchorage is reached, and the mate steps up to the console, miming releasing the brake to let the anchor down. Fifteen fathoms on deck! he calls. Thirty fathoms out of the howse! Thirty fathoms on deck. Brake set two man tight!

Now, anchor chain is strange stuff, and dropping anchor more dangerous an exercise than it first appears. The anchor chain, each black-painted oval link the size of a loaf of bread, runs from a chain locker beneath the bosun's mate's feet through a fat pipe—the howse pipe—onto the deck, where it winds around a capstan, a kind of winch designed to grab the links to feed them in or out, then over the side to one of the two anchors, one hanging off the port bow, the other lodged just below the stem—the very front—of the ship. Aft of the capstan several feet is a console with controls for the capstan. You can also set a brake from the console, a brake belowdecks meant to slow the passage of the chain as the ever increasing weight of what is overboard wants to pull what's still on board with it.

Though water depth on nautical charts is measured today in

feet, anchor chain is measured in an antediluvian unit, the *shot,* which itself is measured in another old-fashioned unit, the *fathom.* One fathom being six feet and each shot of anchor chain measuring fifteen fathoms, one shot is ninety feet long. The chain, Bosun's Mate Allen Lovelace explains, is marked at certain fathoms in color codes so we know how many fathoms we are putting out. Red, white, and blue.

You can "read" the chain by watching the colors at these shot intervals, red at fifteen fathoms, white at thirty fathoms, blue at forty-five fathoms with white links immediately surrounding the marker link to give the eye time to register and record what comes next. The port anchor, Lovelace continues, has twelve shots of chain and the center anchor eight shots. Then, near the very end of eight shots on the center anchor, you will have fifteen fathoms of yellow links; that means caution, you are near the end of your chain.

If you're standing here, he says, gesturing to the console, when you see that yellow, you run. I'm serious. Over there, over here—he points to the port and starboard rails—you just run. At the very end, the last shot, the anchor chain is all red, all red. And there's a saying—red, you're dead! 'Cuz, I'm telling you, when you get to that point with that much anchor chain over the side, there's no way you can stop it.

They say you can set the brake, he continues, his voice as doubtful as his smile. We tried to do that once when I was a sea-man, and it didn't happen. We was actually anchoring in water that was a little bit deeper than we thought it was. The chain just left, went right out! Because it's only attached down there to the chain locker with a shackle, pulled the whole bulkhead out with it. The chief says, get the hell out of here, so we took off! The cap-tain got fired. Lovelace shakes his head at the memory, smiling ruefully, white teeth gleaming against his dark skin. That was twelve years ago, and it wasn't a pretty sight, he finishes.

On the aft missile deck, Torpedoman First Class Valarie Williams is tending to the rack of three compressed air torpedo tubes that make up one of the two torpedo launchers that are her charge. As

Torpedoman's Mate First Class Valarie Williams next to her "babies," the port triple-tube torpedo launcher. "When you rub the nose of a torpedo," Valarie says, "it's so soft. I tell the new guys, first time it's free, then you have to pay me to touch it!"

the only torpedoman on the ship, Valarie is responsible for the maintenance and upkeep of the tubes and the torpedos she fondly refers to as her "babies," her "fish." These big machines, she says, are outside in the weather, and they are going to lock up. If they got gears, you know all that has to be oiled and greased. I have to train on how to load my fish on this ship, too—I've never done it before—because as weaponeers, we always move our own stuff. I'll have to store the weapons, and then, if we go into hot spots, I'll have to load the tubes, then stand by and stand ready.

It took a lot of school to train me to be a torpedoman, Valarie says, but when I got here, I got picked to be mess deck master of arms. They told me, the crew is tired and depressed and we want you to keep the morale up. So I get up at 4 A.M., tired as anyone else. I go into the crew's mess, and I'm in charge of nineteen other people maybe one year in the Navy or less, and I got to get them boosted, spirits high, and we go in and set up breakfast. I make 'em shine all the napkin dispensers.

The crew comes in for breakfast at 6 A.M., and I got to get *them* happy and boosted up, their energy going. We clean up and get inspected at 10 A.M., then we go through lunch. . . . In the meantime people are [injuring themselves] . . . getting cut, or they hit their legs and I've got ice packs on them. They think I'm a battle dress station. I see 'em, blood running down, ask, what happened to you? Come here! All day long. They call me the mess deck mom because I'm always asking, what *were* you doing? You stuck your hand down the drain and got cut on sheet metal? Are you nuts?

Then, she continues, there's a fire drill at three o'clock, and the people are coming into the mess deck, and all they know is, someone just told them where to show up. God! So I'm trying to help them get dressed, gloves on, OBAs [oxygen breathing apparatus] on, and send them out the door to go fight that fire. Because if this ship catches on fire, *these* are the guys that's gonna save my life! So I try to train 'em and dress 'em and they got their clothes hanging out and I put their gloves on for 'em. Then they come back, they got water dripping off them, they get sweat on the tables, and I'm like, just get off my mess decks! And then it's dinnertime!

I've built torpedoes, Valarie says, and I know everything about keeping them happy, but I've never fired one, never seen one go out in the water.

Until recently Valarie's story was not an unusual one, because women at sea were confined to serving aboard noncombatants— auxiliary and supply ships—before 1994. Valarie has been at war in the Persian Gulf, loaded out all manner of ammunition and weaponry to SEAL teams from a supply ship, her first sea duty, yet not had the chance to practice the weaponeering skills of which she is so proud. Having women on fighting ships is still pretty new to the Navy, and of all of the thirty-seven women aboard the *Cook*, none has served on a combatant before. It is the captain's first command with women as well; and for a lot of the men who've been to sea only with their own, there has been a period of adjustment, too.

On a Friday night a few weeks later, Valarie gathers with her two rackmates, Sonar Tech Lori Naron and Electronics Tech Beth Holz, in the basement bar of J.R. Maxwell's in downtown Bath for a not-so-quiet beer after a week in which they've all put in their share of fifteen- to twenty-hour days.

Valarie has traded in her workday Navy blues for jeans and a bright red shirt, her face is lightly made up, and a gold necklace in the shape of a half-moon hangs around her neck. Her hair, which is usually escaping in wisps from underneath her uniform cap, has been liberated for the evening, long shiny auburn tresses of which she is duly proud. She is in the midst of a story, and her china blue eyes flash and sparkle, her hands, the fingernails scarlet, flying around to illustrate the high points as the words emerge in a pleasant South Carolina drawl.

I never owned a pair of jeans until I joined the Navy, she is saying. I always wore dresses and high heels. In fact, I was a runway model. I was twenty when I got married, twenty-one when I had my first baby, twenty-three when my husband [now ex] joined the Navy without telling me, and it made me so mad I went to his recruiter and signed up. Boot camp was like, my God what did I do? I have two girls, seven and fourteen. My baby's been in day care since she was six weeks old. My last command, I spent every day with them, and four years before that I was in and out to sea for short periods. This is a combatant command, and I've got six months . . . I've never been away from them for six months. I don't know how I'll do it. I miss them right now. But I do [this] for them. I have health insurance, I don't have to depend on anybody, and the Navy takes care of my family when I'm here.

I always tried to put up that front, Valarie says, to be tough, because I'm so . . . feminine. So I'm a weaponeer. On my last ship we had to run around with guns, put people down on the deck, put your knee in their chest, make 'em cry. It's wonderful.

I used to take care of the SEAL teams, give 'em their ammo. I knew what they needed, and I got it for them. We kind of bonded. They said one day, you want to come out here with us, hook up to a SPIES line [Special Patrol Insertion and Extraction System; a rope hanging from a helicopter that SEALS clip themselves onto for getting into and out of confined landing zones quickly]. I said, sure, I'll do it, I'm tough. And they hooked me up. All I had was a little life vest, no parachute, they hooked me up to this rope, five of us at intervals, and we walk until we can't walk anymore because the helicopter the rope's attached to takes off! There are five of us on the line, dangling, way up in the air, and the cars are this big! Valarie puts thumb and forefinger together until they almost touch.

I feel like I'm going to puke. I thought, if I puke, those guys below me are going to get it. And then that would be embarrassing because—because I'm tough! And, you know, I'm the only girl up here, too. I didn't puke!

And you've got to open your arms up, and your legs, sort of spread-eagle, so you don't spin around. Because once *you* start spinning, everybody below you starts spinning, too. I was so scared, but it was exhilarating. Soooo scared, but it was just wonderful. I won't even go on a roller coaster because I'm so afraid of heights. But I was doing this to prove to them that I wasn't scared. I did it. I couldn't believe it, but I did it.

Lori and Beth nod in understanding at this need to prove yourself. I'm the only female in my division, Lori says, and I have to work three times harder. Every day I go through something, every day I hear fifty times that women shouldn't be in the Navy. That doesn't stop them from giving me the job, but I have to do three jobs in the time that somebody else does one in order to get recognized. The guy in my work center who is over me, I know how he feels and he knows how I feel. So we hit that wall every day. But he's coming around. It's hard for him because he's never worked with women before. I can already see a change in the guys, they are starting to trust me. The only thing is, if *I* can't do something, it's because I'm a girl. When a *guy* can't do something, it's just because he couldn't figure it out.

Lori is quiet, shy, and, in her early twenties, lacks the seeming natural air of authority of her two more experienced rackmates. Though she talks softly, she is far from demure, with a powerful physique and strong Black Irish features that betray her roots. She came in the Navy wanting to join the elite Explosive Ordnance Divers. She ended up on Coronado Island off San Diego, where they train the SEALS, waiting for her orders. I was the only woman there, she says. Four o'clock in the morning somebody would come into my room bouncing on my rack and yelling at me that we had to go run seven miles. It was great, but it sucked all at the same time because I was always having to prove myself.

I could not do *one* pull-up when I got there, and I felt so inferior to all those guys. For them it was so easy. I knew right then that if I was going to be able to do what they do, if I wanted to be like them, that I had to prove myself, because they already thought I shouldn't be there. By the time I left there, I could do ten sets of twelve pull-ups.

To get on a ship like this, Beth jumps in, there's a screening process, and it's different for men and women. It kind of amazes me still, how long have we been in the Navy? It's like you are trying to break in a club, a boy's club. If they just put bags over everybody's head, took away the gender issue, and said, do you qualify, and any time you qualified, they didn't look to see if you were male or female, then you'd have the most qualified crew out there.

At the first, deceptive glance, Beth looks and sounds like she made a wrong turn off a Florida beach at spring break a couple years ago and somehow ended up in a recruiter's office. Her voice has a lazy Florida lilt, and she is quick to laugh, her smile emphasizing the sun lines at the corners of her eyes. She's slim and moves with grace and speed, her short, straight, dark-brown hair, small ears, and very clear blue eyes adding to the impression of compact energy.

After nine years in the Navy, she is an ET-1, an electronics technician first class. In reality she has a key job as the *Donald Cook*'s local area network administrator, a job of nightmare complexity and long hours on a ship that relies so heavily on the smooth operation and interaction of hundreds and hundreds of computers large and small. She admits to her share of eighteen-, even twenty-hour days in the past few weeks, but this doesn't seem to have dampened her excitement about her job, her ship, or her place in the Navy. She is very much driven by specific goals and is seriously considering applying for an officer program.

It's not hard to see her as a "mustang," what the Navy calls those chosen few prior-enlisted at the top of their ranks with spotless records and usually ten to twelve years of priceless shipboard experience who make the huge leap to the officer corps through several different, very selective training programs. Such officers are called LDOs, limited duty officers, because they cannot rise above a certain rank and will never be offered, for example, the command of a ship.

There is, Beth says, a lot of respect for mustangs in the Navy. I think most enlisted will say, the best officer is a prior-enlisted because you've been there, you've done that, you've taken all the crap, been through all the stress. There are a lot of LDOs on this ship because the Navy needs their experience. In fact, you couldn't put a junior officer right out of school in this level of production right here. They need somebody with a technical background, the years of experience, who knows the griping, knows how to cut corners, knows how to get around this, runs into walls and gets over them.

Beth doesn't just think the Navy is better off for having women, she knows it. And it's not only the very real contribution they make on the job, but the effect their presence has on the functioning of the ship as a whole. Having women on ships, Beth asserts simply, definitely puts a softer edge on things. A bunch of guys on a ship all alone is just a bunch of guys.

They're pigs! Valarie says. They have no manners . . .

And with women, Beth continues, the manners come back and all these other good things that don't take away from their ability to do anything, other than the civilization level comes up a bit.

No matter the level of civilization aboard ship, what is certain is that daily life for women in the Navy is just not that easy, especially if they are ambitious or want to compete, like Valarie, in traditionally male bastions like weaponeering. There is disapproval from another source as well, as Valarie explains.

My last ship, she says, had five hundred women on board, and when we came back into port, you're getting off the ship, you've

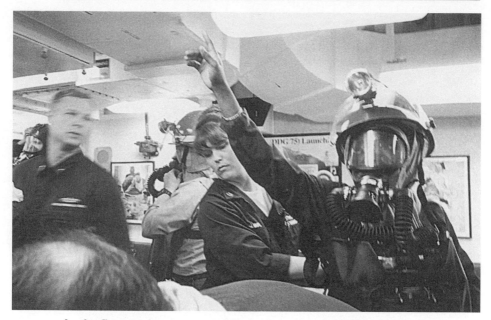

As the fire attack team assembles in the crew mess, TM-1 Valarie Williams helps a firefighter suit up.

got your sea bag, and the guy's wives are at the gate. They're calling us whores, sluts, other names. What I told the other women I was with, I told them just don't say anything, because you got to feel what they feel. You're two weeks out in the sea with their husbands. People are ignorant. In fact, there's a woman writes a column for a newspaper in Norfolk. She writes about how women shouldn't be in the Navy and how we're all sleazy . . . and that's a Navy wife!

At about three o'clock in the afternoon, five, sometimes six days a week, the *Donald Cook* experiences a potentially ship-killing disaster involving fire. Though the crew knows there will be a drill, they don't know where the fire will be located. Some of the sailors new to the ship don't even know their way around yet. When they hear that first, urgent announcement, however, they run to the muster area in the main crew mess, grabbing equipment from racks as they go.

Earlier, one of the senior officers has made his way to the site of the exercise and illuminated the "torch," a fake fire that glows ominously behind a thicket of pipes at the forward end of the main engine room, which is three decks down and just a little forward of the middle of the ship. At the same time, smoke generators begin to churn out clouds of dense black smoke that quickly fill the confined space.

The main crew mess deck is nearly empty but for Valarie, who's standing by for her role as dresser, and a few sailors drinking cups of lemonade or coffee around the room. This relative calm is suddenly and brutally broken by the PA announcement.

Major fuel leak! Repeat, major fuel leak in main engine room.

Immediately all hell breaks loose in the mess as about forty people, most with their arms full of firefighting gear, pour through every door. While small groups set up communications and muster points in one corner of the room, two fire attack teams of twelve sailors each begin shedding their uniform boots and pulling on knee-high fireproof boots in the crowded spaces between the cafeteria tables. The silence is obliterated by the clank of the oxygen breathing apparatus on tabletops and nervous shouts for help. Who's got a canister? A canister! A kelly-green oxygen canister that fits into the front of the OBA goes sailing from one side of the mess to the other.

Fire! Fire! Fire! Fire in Main One! Location of the leak: discharge side. Mechanically and electrically isolate Main One with the exception of lighting, firefighting equipment, and ventilation! Set primary and secondary fire parties for Main One! Primary forward, Frame 174; primary aft, Frame 220. Secondary forward, Frame 126; secondary aft, Frame 254. Set smoke boundaries for Main One! Engineering duty officer has assumed control of the firefighting efforts in Main One.

Here! Here! another sailor gestures, and Valarie gamely wades in, helping him put his arms through the OBA, tightening the cinches at the back and front, fitting the fireproof hood and neckpiece that goes under the face mask. All around the room, half-dressed sailors stumble and reel about, the slow ones still unknotting recalcitrant bootlaces while others hold their arms up like surgeons waiting for gloves, though theirs are heavy fireproof mitts rather than latex. Let's go! Let's go! Let's go! the few officers present urge, knowing that every second counts. Who's got a canister! Need a canister!

Fire! Fire! Fire! Fire in Main One! Fire in Main One is out of control, evacuating space.

A whole communication setup has appeared on one mess table, and reports on the action begin to flow to and from the CCS, the Central Control Station, from where all of the ship's major systems can be monitored and controlled. After a few minutes most of the primary fire attack team has fully suited up, gathering around the doorway nearest a hatch leading belowdecks, a sea of bobbing cherry-red helmets with headlamps attached. They're waiting for word on the status of the fire, which, because it is in a main engineering space, can first be fought with the halon firefighting system in place there.

Halon is a chemical that, when released into a closed space, binds with available atmospheric oxygen to deprive the fire of fuel. The most critical spaces where fire is likely because of the presence of fuel, the engine rooms and turbine-driven electrical generator spaces, are all equipped with this system. Halon is lethal to a fire, but also to people. Anyone caught in a closed space when the halon is lit off will suffocate in a matter of seconds. As well, the firefighters must wait for up to fifteen minutes after halon light-off for the air to clear enough to breathe.

Section 1 fire marshall, bypass primary halon, light off secondary halon. Indication that primary halon—is bad!

Halon's bad, you guys gotta go, the section leader calls. He is Senior Chief C. J. Oakley, thin and intense and with the forceful voice and presence necessary to impose some kind of order onto the churning mass of half-kitted-out teenagers milling around all around him like a herd of panicked cattle.

Come on! another officer yells in frustration. Halon's bad. Getcherhelmetonwhereareyourglovesjesus! Go, go, go! he says, nearly pushing the heavily bundled men and women out the door. Fire Team One, where's your other guy? Get out there! Valarie appears here there and everywhere throughout the packed mess deck, settling a hood over a young woman's shoulders, cinching straps of OBA behind a sailor's back, tugging on gloves, and finding helmets that have gone astray.

Halon is bad in Main One. Halon is bad in Main One.

The primary fire attack team musters out, and the secondary team waits on deck. Talk to me, talk to me, someone is calling over the radio. Fire team's communications crackle in background, muffled by helmets and face masks as they make their way forward and down to the main engine room. Attack Team Two roll out! Oakley urges. Go, go! Go set your boundary! Go!

With a fire on a ship, Oakley says, there's nowhere to go. It's hot, and it's completely dark. You got to cut all power to the space, so they are shutting everything down in Main Engine Room One. When the attack team goes in, it's just pitch-black. All you can see is the fire, and it's of course producing black smoke. So besides all the heat, you can only see about a foot in front of your face. It's bad enough to practice it, but the real thing must be terrifying. That's why we send out boundarymen, to try to isolate the fire in case it gets out of control. They go to outlying spaces around the fire area. We do it from top, bottom, the sides, and the face. They mark the fire, if a bulkhead gets hot or the paint starts cracking, you know it's comin' at you and you can move your boundary back. You may not be able to put it out, but you can contain it to some space.

The fire attack teams will first have to pinpoint the location of the fire exactly enough to know where the nearest fire hose will be. Then they will unroll the hose, charge it with water, and "douse" the fire. They may also use fire extinguishers and fire-fighting foam. Keeping in mind all of the compartment doors they have to pass through on the way, this can take more time than one might think.

From one end of the ship to the other, every pressurized air-lock—two heavy oval compartment doors with enough space between for four people to crowd in—is closed, dogged down to seal it tight, and manned. These airlocks divide the ship into separately ventilated zones making up the ship's Collective Protection System, so that smoke and toxic fumes in one space don't get sucked into adjoining spaces through a common ventilation system. Sealing up the ship and controlling access also makes keeping track of who is where much easier. Keeping the doors closed therefore becomes of primary importance, such that any sailor passing unauthorized through one during a fire drill automatically becomes a casualty who has to be carried out of the space to the nearest dressing station, which absorbs manpower and complicates the drill.

With both teams out, the mess deck is almost empty after the rush. Abandoned boots and clothes litter the tables and floor as if a herd of unruly children has passed through.

Boundaryman Two, Boundaryman Eight, make boundary report to CCS! Mechanical isolation complete in Main One! Attack team has permission to enter the space after the activation of the bilge sprinkler for two minutes.

Sprinkler activated! crackles over radio. They're fighting the fire, Valarie observes. I'm happy helping with all this craziness 'cuz they have to get it right if they're going to save the ship.

Electrical isolation is complete in Main One. Electrical isolation is complete in Main One.

Status and damage reports pass back and forth from belowdecks to crew mess to Central Control Station and back. Minutes pass, and finally the all clear sounds.

Fire in Main One is out! Reflash watch is set! Secure from drill. Restore all casualties, restore all gear. Reset electrical and mechanical isolation. Another quality drill, good effort by the duty sections. Nice and quick, start to finish in about thirty-two minutes today.

The first members of the primary fire attack team straggle in, sweaty and exhausted. Man, says one, we ran all over hell and back. I am hot. Valarie helps tug off OBAs and offers cups of cold water to the sweat-soaked sailors.

Attack Team, you did a good job! Senior Chief Oakley tells the assembled fire teams. This, he says, is maybe only the tenth time they've actually sent the fire teams out on their own to see what they were going to do and how they were going to react. Today was the first day they started throwing other variables in. They lit off the halon, and two minutes later they said the halon's no good, what're you going to do? They're getting better and better. Put the fire out and then it reflashes. It could kill off a whole hose team. Say you got a missile hit, and you don't have a fire main on the starboard side, no pressure in your hoses. You got to be ready.

As he speaks, the sailors gradually disperse, the clatter of helmets and OBAs on hard plastic and steel diminishes, Valarie makes her way around the room, rescuing forgotten gloves, hoods, and oxygen canisters from under tables and stowing them neatly. She wipes down tabletops and shoos people out of the way. It's nearly dinnertime, and she's got to have the mess deck shipshape for the hungry hordes soon to descend on her clean mess.

And she must be ready to greet each and every one of them with a smile and, perhaps, a gentle reminder to remove their hats. This is the crew mess decks, she is the master-at-arms, and this is the Navy, where tradition counts for more than in most places, and where you always take off your hat before sitting down to a meal out of respect for all of the Navy dead who came before you.

10

CHARLIE TRIAL

Remember that you are on a moving platform . . . "One hand for yourself and one for the ship" is an old but still very wise practice. . . . Do not open, or close, any valves, flip any switches, push any buttons, or play with any system controls.

—*USS* Donald Cook *Charlie Trial "Welcome Aboard" pamphlet*

The bridge of the *Donald Cook* is a magical and intimate place, especially late at night when the pace of the trial testing slows and all eyes turn towards the sea, alert for the particular dangers that darkness brings. Shaped like a flattened horseshoe, its slightly curving outer face lined with thick glass windows from shoulder height to ceiling, the bridge gives a commanding view in all directions but directly aft, where the turbine intakes and uptakes—the stacks— block the sightline. It is by far the quietest place on the ship, five floors above the main deck, insulated by layers of steel and Kevlar from the crowded crew spaces, narrow corridors, and noisy machinery rooms below.

In the center at the rear of the twenty-foot by twenty-foot space is the steering console, looking like nothing so much as a video game. The helmsman, Rick Townsend, sits in a padded chair in front of the controls. He grips a small black steering wheel, his eyes on the red glow of a readout on the panel above the controls that tells him his heading and relays numerical information about the positions of the ship's two rudders and the functioning of the rudder control systems.

To his right is a chrome-handled throttle similar to that found on a jet airliner. Above the throttle are red-lit indicators showing the status and power levels of each of the four gas turbines, two of which drive each prop shaft. From a neutral STOP setting, pushing the throttle forward ramps up the turbines through AHEAD 1/3, AHEAD 2/3, STANDARD, FULL, and FLANK (maximum speed). Pulling it back from STOP ramps the power up again—BACK 1/3, BACK 2/3, BACK FULL—but in reverse. On these ships, movement in reverse is accomplished not by changing the direction of rotation of the prop shaft—which is impossible—but by changing the pitch of each blade of the propeller.

Joe Pinette, a compact, fit former merchant mariner in ball cap and sneakers, hunches over a chart illuminated weakly by a red light. The third mate, Pinette is acting as the navigator, plotting the ship's course due east out into the Gulf of Maine. It is early evening, and the ship has come about thirty nautical miles since leaving the BIW pier twelve miles up the Kennebec at 4 P.M., the afternoon passing in a blur of tests and inspections that make up the last of three trials BIW must put the ship through to certify its worthiness to the Navy.

Throughout the first two trials in earlier weeks, Alpha and Bravo, every system—propulsion, power, weapons, fire and damage control—has been brought on-line and tested. Somewhere off Virginia two weeks earlier, the five-inch gun on the foredeck boomed out test rounds, the Vertical Launch System fired its first missiles, the two Phalanx CIWS (Close-In Weapons Systems) 20-mm cannons filled the air with a curtain of depleted uranium projectiles at three thousand rounds per second, and torpedos sped away from the ship towards a target. The ASW (anti-submarine warfare) systems simulated sonar detection and tracking undersea, while the Aegis guided missile systems located, locked on, and tracked simulated overhead threats.

Charlie Trial, BIW's third and final hurdle, is the brief but intense final certification and acceptance by INSURV, or the Naval Board of Survey and Inspection, an independent, congressionally mandated panel that puts its final imprimatur on the ship.

The trial is also a contract milestone, a successful one culminating, certainly, in handshakes all around; but, more important, in a large check from the Navy to the yard.

Charlie Trial finds the ship at an odd point in its evolution. Its ownership has not yet been transferred to the Navy, so it is manned largely by yard workers, who handle the lines during docking, and whose burly bodies are crowded into the machinery and control spaces behind consoles, monitoring everything from turbines to firefighting pump systems, and populating the bridge. All of the ship's defensive and offensive weapons and communications systems are up and running, too, and these are operated by a mixed crew of BIW employees and manufacturers' representatives, with Navy personnel observing and participating as well.

The INSURV presence is everywhere, of course, active-duty officers in summer khakis overseeing the survey with the help of belowdecks minions in gray Nomex jumpsuits, the oval INSURV logo picked out in red letters on a white background, bringing to mind 1950s gas station attendants.

Working with INSURV and BIW are representatives of the Supervisor of Shipbuilding, Conversion, and Repair in Bath (SUPSHIP). SUPSHIP Bath is an important and prestigious naval command, this one led by Captain Ralph Staples, who guides a score of senior Navy personnel and four hundred qualified civilian inspectors in overseeing, testing, and approving thousands of different matters during construction, from ultrasound scans of critical welds to the correct application of paint on the hull. SUPSHIP is also a financial and budgetary watchdog, dispersing more than a billion dollars every year in contract money at major points in each ship's construction.

Also aboard are the *Donald Cook*'s officers and department heads as well as a more select group of bluejackets—men and women from the lower ranks. They are easily identifiable not only by their light-blue button-down shirts over flared blue denim trousers, but because they are so young, with their hair shaved close to the head on the sides and grown perhaps an inch or two on top, giving them a distinctly cool look. They are there to accelerate the process of becoming one with their ship, to know its whys and hows, to begin to form the ties with it and with one another that will allow them to function as a single unit once they've left the dock on their first voyage.

Charlie Trial formally began two days earlier at eight in the morning dockside, with weapons systems inspections the first day, and more mundane demonstrations—inflatable life raft stowage, damage control and fire station inspection, laundry and galley equipment—the next. This day, when the ship goes to sea, is what everyone waits for with sweet anticipation, an excitement evident in the wide grins and backslapping and howyadoins amid the intense preparation.

The pilot who will take the ship down the Kennebec River and out to the test area in the Gulf of Maine has been on board for several hours. He is Captain Earl Walker, and his Coast Guard–issued license identifies him as a "Master of U.S. Steam or Motor Vessels of any Gross Tons, Radar Observer, and First Class Pilot" in all significant harbor approaches from Bath and Portland south, through Cape Cod Canal, Block Island Sound, New Haven Harbor, Long Island Sound, New York Harbor, down to Perth Amboy and Newark Bay, New Jersey.

Diminutive of stature, a bit rotund and white-haired at sixty-three, he's got a face that has seen some weather, a demeanor that radiates confidence and good cheer as he moves energetically about the bridge in neat khakis, a dark blue USS *Donald Cook* baseball cap on his head. Though a civilian engaged along with the three mates by the yard for this trip, Walker has absolute authority. He gives the orders and everyone else—uniformed Navy, SUPSHIP, contractors, or yard employees—follows them. There just aren't that many Coast Guard–licensed Masters who are also Kennebec River and Portland Harbor pilots, and certainly none with Captain Walker's more than fifteen years' experience moving BIW ships from Bath to Portland and out into the Atlantic for trials.

Many things have to happen before the *Donald Cook* pulls away from the pier. Along with the scores of machinery, electronics, and communications checks, things are happening outside the ship as well. A dive team has a man in the water checking to see that no wayward line or river trash is fouling the props. The safety boat is shooing away curious spectator boats—including one insane shirtless macho man in a canoe who seems intent on coming close enough to lay a hand on the hull.

Two tugs have edged up to hover off the port-side stern and bow. They'll keep the ship snug to the pier until she is well and truly under way. Not only are there weird little eddying currents created by the local geography, but once the ship's turbines are engaged, the props always spin, no matter how slowly. Even with the pitch of the prop blades set at zero thrust, she wants to back up slightly. Finally the *Donald Cook* gives a blast of her horn, her stern swinging out, the bow coming around until she is off, out into the middle of the river.

Up on the bridge, things are intensely quiet. Short radio messages pass between tugs and ship. Small groups of khaki-clad chiefs mutter over charts. Captain Walker does not talk very much, calling out a bearing, swiveling his head from side to side. He ramps up the power so the ship is gliding along at a moderate speed, 9 or 10 knots. Fast enough so that the ship responds to the rudder, but not so fast that he's swamping day cruisers come to play tag, and ripping docks and floats from their moorings all the way down the river.

The first obstacle is the Kennebec River itself, twelve miles of serpentine curves, fast-flowing tidal currents, and creeping ledges and shoals that can reduce the depth of the river channel by several feet in a matter of weeks. Captain Walker knows this river from the bridge to its mouth intimately, so intimately that he can draw it from memory, accurately and to scale, every rock, shoal, and aid to navigation. This critical recall ability—drawing charts from memory—is a condition of his license to move ships of any size on its waters.

If Captain Walker knows her like a lover, he knows that she can be a treacherous lover, indeed. Which is why, a couple days before this trip, he could be found tracing the river channel in the small tug *Kennebec*. In the tug, too, were three people from the *Donald Cook*—Commander McCarthy, the ship's captain, his navigator, Lieutenant James Katin, and their lead quartermaster, who sets up the map teams and oversees the taking of bearings as they go down the river. We try, Captain Walker says, to take a

command officer or a navigator so they have a chance to familiar-
ize themselves with the river. The spots I know are always problems, we'll sound quite heavily.

These spots are numerous, and the first is not long into the
trip, just past Doubling Point, a notoriously sharp bend where the
river jogs east almost at a right angle. Complicating matters is
that, here and elsewhere farther downstream, the channel can be
very narrow.

A mile south of the yard, abeam of the number 31 buoy,
Captain Walker says, is your first point where you're really concerned about water depth. He pinpoints it with a finger on a chart
spread out in his crowded office on Union Wharf in Portland. The
chart shows a shoal protruding into the channel off the western
side. Walker continues, We had the Army Corps of Engineers dig
that out last November. Well, twenty-seven feet is the mean channel depth for the Kennebec, and the Corps will only overdig by
two feet. That river moves an awful lot of silt and an awful lot of
sand on every tide. In two weeks' time, the silt had just filled it
right in again. That's why we only move the ships at high water.
It's not all that unusual to have a spot really closed up tight and
only have twenty-five feet of water over it. We set the draft of the
ships at twenty-nine feet six inches, and the tide gives six, six and a
half feet above that, so you are using every bit of tide.

Doing the simple math reveals that the ship has less than six
feet of leeway above the channel bottom at high tide, two feet in
the shallowest areas. Were the ship to be caught at many points
along the river even three hours after high tide, it would run
aground. At low tide proper, the ship would be sitting in the mud.

Though the day is hot and hazy, the water calm, and no significant weather on the horizon, there is still a good deal of opening
night jitters on the bridge. Captain Walker has piloted this transit
dozens of times, serving as a mate and getting to know the river's
tricks for a handful of years before stepping into the retiring pilot's
shoes. He also knows that one of the things he cannot control is the
weather. In this part of the world, his main adversary is fog, which
can creep in over the trees and cloak the river in a matter of minutes. I won't, Walker says, shaking his head and looking serious,
knowingly move the ship in fog. If I get caught, it's another situation.

Transiting the river in 1986, piloting the USS *Simpson,* a Bath-built guided missile frigate, Captain Walker got well and truly caught. All the way up the river, he says, the tug running along the riverbank ahead is radioing back, *I've got good visibility here, Earl! My* visibility was zero, or, it would open up for a moment and I would see him. Then I would see the fog coming in over the trees and dropping right back in the river. It was like that all the way up the river. If I hadn't entered the river with good visibility, I never would have risked going on.

You have to assess the situation at the time, Walker emphasizes. What's my tide? Current? How are my radars—are they functioning properly? With the new Raytheons, I could do it but I wouldn't *want* to. Because I would have to keep speed on. I couldn't be jacking the speed up and down because I would have to know I was staying on track. As soon as you reduce your speed, the current takes control. And then you're going to be off track. Unless you pick it up instantly on the radar, boy, you are in trouble.

And that's part of piloting, Walker says matter-of-factly. Doing everything the same way every time. So that, when something doesn't look right, you know instantly in the back of your head that it's not right.

Captain Walker rummages in his canvas briefcase and comes up with a packet of photographs. He pulls out a particular photo, taken straight ahead off the bow of a small boat looking to a wooded shoreline. In the water ahead, barely visible, is a cone-shaped float called a *nun,* and behind it, a smaller, barrel-like buoy. A black arrow in magic marker points to a notch in the trees where treeline meets sky, with a bearing added, 135 degrees.

People laugh at me because I carry some photos, he says, chuckling and sort of shrugging his shoulders. *Jesus, this guy is piloting by trees,* they say. It's no joke, it's the truth! I'm not unique in this; pilots do it and have done it for centuries. The captain turns to the photo, explaining that it's the leg of the river past the town of Phippsburg on the western shore, about halfway down the river, where a shoal extends out behind a thumb of land called Squirrel Point.

The way to stay off this shoal, he continues, you see this little V-notch in the trees there, you want to be steering 135 degrees on that notch. Sometimes in the early morning when the sun is coming up, it's difficult to see those buoys at all. But the V-notch in the trees stands right out. That's down-bound. When I come back up the river, I steer 315 degrees on the Phippsburg church spire.

All the way down the river, Captain Walker uses not only cans, buoys, nuns, and other traditional navigational aids, but church spires, light towers, a Coast Guard station, a particular green boathouse, a certain dock, a set of iron rails where someone puts their boat in the water.

The water tells him things, too. Coming downriver against the incoming tide around Doubling Point the ship must make a sharp turn left (east). As the ship begins to turn, the flood current hits the starboard bow and wants to push the boat west, into a shoal, an effect magnified by an eddy current. In this area, Walker says, you can watch the tide line—a bit of foam, a little debris, seaweed—in a line. You tend to come wide unless you drive yourself into that turn, just blow her around it, and I keep my bow just inside that tide line.

Two hours out, the ship has passed down the last of the narrow, rocky banks of the Kennebec below Bath, passed Fort Popham with its sand beach and kids playing and ranks of surfcasting fishermen hopeful for striped bass. Pond Island Light slips by the stern and then, out into Casco Bay, the deck vibrates as the ship's speed moves from a barely perceptible 8 knots up to a cruising speed of about 16 knots.

A short while later, the ship having passed Seguin Island about three miles offshore, her speed falls away to nothing until she is drifting, dead in the water. By this point, Captain Walker says, the engineers are always in a panic, asking, *When are we going to be there? When are we going to be there?* It's not only the scheduled drag shaft and upcoming full power test, but with all the people we have riding and drinking coffee, they are really taxing the system.

The captain is talking about the sewage and wastewater tanks, which are full to brimming and must be pumped, but can only be pumped outside the three-mile limit. After the dead stop, the drag shaft tests begin, in which the engineers shut down the two engines yoked to each shaft in turn, thereby dragging that shaft through the water with the prop spinning free, like a freewheeling bicycle tire. The purpose is to zero out their instruments so that, in half an hour when all four turbines are running straight out at full power, their readings will be true.

Up on the bridge Captain Walker consults with Mark Colby, a mechanical engineer who serves as the official trial coordinator. Both radars show an open sea ahead, a clear seaway for the full power run, which is not only the first, but the most important of the underway tests. Captain Walker confers with Colby, and Colby announces the full power test at 6:15, in just a few minutes.

Captain Walker gives the order, the throttle handle creeps forward, and the deck vibrates and shudders slightly as the ship takes off. The bow begins to arc up out of the water, cleaving a deep groove in the gray-green water as the two enormous propellers tear up the sea behind.

All the way aft, the flat open space of the fantail is dotted with people come to enjoy the show. They emerge from the starboard compartment door off the main deck, legs spread, hands clutched to hats in the buffeting wind, moving slowly across the rolling deck to the rail. Off the back of the ship, the great force of the turning blades throws up a twenty-foot rooster tail, tons and tons of water surging uniformly upward in a column, then falling forward, disintegrating into a jumbled, confused mass of ever changing shapes. The setting sun follows the ship, its red-gold light firing the liquid forms from behind like candlelight on crystal, a mesmerizing, ever renewing spectacle.

This is the ship at full power, 32 knots, or almost 40 miles an hour. Is it the fastest she'll go? Here things get murky because, for understandable reasons, the Navy would rather keep silent. Full power is probably not, however, the absolute top speed of the ship, for there are two throttle settings even higher—FLANK and EMERGENCY FLANK. In general, FLANK delivers maximum power at a sustainable speed, a few knots faster than FULL, just at the outer edge of the normal operating envelope for which the hull and turbines are designed.

During EMERGENCY FLANK operation, the turbine governors that normally limit the engine's maximum power output are electronically disconnected, allowing the kind of do-or-die unsustainable surge of power necessary only in extreme circumstances, such as man overboard or last-ditch maneuvering. Exactly how much higher this theoretical maximum is than 32 knots, the Navy also prefers to keep to itself.

At cruising speed in a relatively calm sea, the ship hardly seems to be moving at all, especially in open water with no fixed visual reference points against which to track your position. Likewise, up on the bridge, the ship's forward speed is almost impossible to gauge without reading the red LED speed indicator flickering up and down, and keeping an eye on the height of the bow wave.

At full power, what *is* noticeable is that the ship feels more like it is flying—moving over the water rather than through it. Given that the source of its power is four General Electric 25,000 hp gas turbines, the same engines that make a DC-9 fly, this makes sense. The just-audible telltale whine of turbine fans certainly adds to the effect, as does the reek of exhaust, the smell of the airport, though neither has anything to do with propulsion. They're produced by the aft Allison turbine generator, which runs on JP-5 jet fuel, while both engine rooms are far belowdecks, the four GE 25,000 hp gas turbines presently driving the ship forward consuming their 100 gallons of Number 2 fuel oil a minute without sigh or hiccup.

Inside, the air is heavy, not with a reek but with an aroma, that of dinner being prepared for four hundred in the main galley next to the crew's mess: rare prime rib, baked potatoes, broccoli, rolls, and strawberry shortcake for dessert. A long line snakes down the already narrow main starboard passage, veering abruptly towards the center of the ship and the galley serving area, cooks on one side slinging the courses onto comparmentalized plastic trays, and hungry men on the other, sliding their trays down the stainless steel rails and talking shop.

Once the trays are loaded, two steps aft and the crew's mess opens up, a cheerful place of dark blue linoleum floors flecked with white, blue chairs bolted to fake-woodgrain tables in four- or six-man groups. One side is taken up with industrial-sized coffee urns, lemonade and red Kool-Aid machines, and—an unexpected luxury that will prove increasingly popular as the night advances—a soft ice cream dispenser.

■ ■ ■

Up on the bridge two hours later, it is nearly dark, the ship moving imperceptibly forward through small swells. Two men stand at the windows to port and starboard on a four-hour watch, never turning their backs to the sea, occasionally reaching for high-powered binoculars. Each has an elevated, comfortable chair at his back, but both seem to prefer to stand. They hardly move, do not speak, and are rarely spoken to, but look out ahead and to the sides, on the watch for the wooden fishing boat or sailing yacht coming within range unnoticed by the radar. The *Donald Cook*, at 8,000 tons, would sink a small boat even by passing too close to her.

The ambit of the Navy's test area runs north towards Nova Scotia, and south and east far enough to almost touch the north-western tip of Georges Bank off Cape Cod. Particularly in the spring of the year, says Captain Walker, but also this time of year, the white whale is mating and calving down in here off the east side of the Cape. The Navy is very, very concerned about hitting one, so they stand a watch, a whale watch, even on trials, in that whole area. Even though my lookouts are keeping an eye for them, they will send their people out to watch during the day and with visibility. One of the things they have in front of them is a folder, so they know whether they are seeing a minke, a hump-back, a finback, or a North Atlantic right whale. We know they are out there, so we try to avoid them.

It is quiet in the night air, the only sound the faint hiss of air conditioning, the clang of compartment doors being locked down nearby, Wally Pitcher's laconic "Bridge," as he answers the phone. A BIW engineer who does testing and evaluation as part of the Hull, Mechanical, and Electrical Department, Wally's been part of the trial team for more than ten years.

At the moment, he's dealing with someone responsible for checking all the deck lighting, agreeing with him in his dry way that, yes, having all the spots on, illuminating the fo'c'sle like a Christmas tree, might interfere with the upcoming Darken Ship tests. Somewhere an alarm goes off, a disembodied voice from a speaker informs everyone that a toilet is overflowing on 2 Level belowdecks, a bluejacket comes onto the bridge, sirring the third

The *Donald Cook* executing a high-speed turn in the Gulf of Maine during
Charlie Trial, July 1998.

mate and informing him that they're going to have to take up a
corner of the floor covering to check an electrical line.

Captain Walker is consulting with the navigator, who's plotting
a more northerly course to take the ship into water deep enough
for a later test of the Nixie, a device towed far behind the ship
that electronically and magnetically misrepresents itself to enemy
torpedoes and mines as an actual ship.

Now the ship is about ninety nautical miles out, and the captain gives the order to begin the test events known prosaically as
the "ahead steering demo" and "astern steering demo."

"Before the full power ahead steering test," BIW's Charlie
Trial Agenda booklet advises, "the word shall be passed 'All hands
stand by for high speed turns.' This means the ship will be experi-
encing extreme rolls and erratic movements. All personnel shall
secure loose gear in their area. . . . Persons shall hold on to the
nearest secure object until the word is passed to secure from high
speed turns. Do not underestimate the ship's movement during

these maneuvers." All the weather decks—any space open to the air—have been inspected for loose equipment, anything that might break free in a sudden heavy sea or the high winds that full speed generates.

In reality, a few minutes later everyone on the bridge has got at least one hand overhead, grasping the copper cables strung at strategic intervals for just that purpose as the ship veers hard to port and hard to starboard at full speed. Because the higher you are off the water in a ship—and the bridge is about fifty feet up—the greater the degree of roll, at 32 knots hard to port the sudden pitch of the deck underfoot can send you sprawling. As the helmsman throws the rudders over to hard to starboard, the flat line of ocean horizon out the windows tilts sickeningly.

The same wide S-turns are repeated at BACK FULL, followed by a maneuver meant to mimic collision avoidance, the *crash back*. At Captain Walker's orders, the helmsman grips the throttle and moves it from FULL AHEAD to BACK FULL with no stop. Beneath your feet, the deck judders as the reduction gears, which act like a transmission, step the rpm of the spinning prop shafts up and down while the prop blades rotate to change the ship's direction. On the bridge, the sensation is one of going from acceleration forward, to less acceleration, to going very fast—20 knots—backwards. Out the windows, the water ahead changes abruptly from whitecapped waves to oily flat calm as the ship is suddenly reversing through its own wake. To any ship observing visually or on radar, this furious ballet must appear, if not erratic, then downright bizarre.

While Captain Walker directs the maneuvers, half his attention is always on the "67," as it's called, a small radar screen on the port side, which has, as he says, a range of ninety-nine miles but-don't-you-believe-it! and which is used for picking up far-off contacts in open ocean. Up front in the exact center of the bridge is a larger radar screen, the "64," a close-in navigational radar that sweeps for shipping traffic inside of twelve miles. To keep an eye on both screens at once, the captain often stands on a box just to the right of the steering console, his eyes swiveling from one to the other in constant vigilance.

He's known as an able and cautious seamen. A unique individual, the captain, just a great, great guy, says Captain Staples, high praise from a career Navy man. Walker's fundamental concern is the ship and his men, not BIW's production schedule, and any damage to this billion-dollar machine would cost him his career at the very least.

This is why he's not afraid to hold the ship at the mouth of the Kennebec—or anywhere else he can safely anchor—for a whole day in thick fog, even though the pressure from above over the lost production time is crushing. There's already one Walker's Point down near Kennebunk, he points out. I don't want another named after me up on the Kennebec.

This was the case with the first DDG, the *Arleigh Burke*. At launch in 1990, it was nearly two years behind schedule and had cost more than twice the initial estimates. Its birthing was accompanied by various technical glitches, a three-month strike at the yard, and all of the normal problems that always crop up in the first ship of a new class.

Added to this was enormous pressure from above to have the ship ready for the first transit as soon as possible. The venerable "Thirty-one Knot" Burke was even then eighty-eight years old, and the press event surrounding his presence on board—a legendary man on the maiden voyage of a much-ballyhooed ship bearing his name—was meant to go some way to signal to everyone, especially Congress, that everything was on target and according to plan.

Captain Walker was the port pilot and Master, and he remembers it quite clearly. The day was just not a good day, he says. It was raining, it was foggy. But we had the admiral and Mrs. Burke. Word came up at the yard that we had a weather window, and that's not to say we had it all the way downriver. With the time it took to get them on board and under way, conditions were changing even as we turned the corner of Doubling Point. By Doubling Point Light, we came down the reach, and I muttered to myself, *Oh, shit!* The secretary of the Navy was standing beside me, and he says, *What's the matter?* I said, *Well, that's fog up there coming over the trees.*

I don't think, Walker adds wryly, that he caught the implication. It was a little embarrassing because Admiral Burke had said we've got such good radar on this ship, we can go just anytime.

Well, the Navy 67 radars just aren't good for navigating. You could just barely tell where you were with them. So we continued downriver, and when I saw the buoy down at Indian Point, I said, we are going to anchor right here. There was no way I was going to try to make the turn at the next buoy in fog with lousy radars. Now, with the radars I have today, the Raytheon, jeez, you just almost could drive nails with that.

So, Walker finishes, the seas were . . . He shakes his head at the memory. It was a *terrible, terrible* trip. We had more people sick and unable to get out of their bunks, just hanging on to the thirty-gallon pails and puking. A terrible trip.

He tells a story about his days as a first mate on the SS *Leman*, a United Fruit Company ship running chilled and frozen fruit products in the South China Sea to Vietnam. We were running on dead reckoning, he begins, it was northeast of the entrance to the river up to Saigon. I looked at the chart and, son of a gun, there was a shoal marked ahead. I put my finger on the chart and right there it says, Walker and Paine Bank. Well, Billy Paine was my mate on my run to Monhegan Island. . . . But I'm not superstitious! So my relief comes and I tell him, no, I'm not going to let you relieve me if you take that course. So he goes to the captain, says, *Captain, Walker says he won't let me relieve if I don't change course!* The captain thought I was . . .

He pauses, and the three or four guys who have drifted over in the course of his telling wait expectantly. Well, I changed course, Walker finishes, and we never did go over that shoal. So, no, I'm not superstitious. But why push your luck! As the others laugh at this oft-heard bit of sea lore, the captain adds a less comical conclusion. Any time you presume something, he says, his face dead serious, it bites you in the ass. And any time you take something for granted, it bites you in the ass, too.

The ship is now passing near Three Dory Ridge, a shallow upthrust where fishermen congregate, and therefore a place for vigilance. He confirms for the navigator that Jordan Basin, northeast some miles, where 100-fathom water is the norm, will be the destination for the coming Nixie test.

As he speaks his eyes take in the whole bridge, moving from the face of the "67" on his right, glowing green as its arm sweeps the horizon. He wanders over to the "64," pressing buttons to eliminate the surface interference caused by wave action known as *sea return*. Set into an overhead panel above the windows on the port and starboard sides are red LED readouts that flicker constantly, clocking heading, speed, and the rpm and pitch of each propeller, registering the most minute of changes. Wally calls him over to discuss the next day's schedule, and his place is taken by Captain Staples, the actual supervisor at the Supervisor of Shipbuilding.

This is a bittersweet trip for the SUPSHIP, for he is retiring in several weeks after twenty-seven years in the Navy, moving to San Diego to take a good job as a systems engineer with a major engineering consulting firm. Looking at him gripping the overhead cable with both hands, feet wide apart, a wistful smile on his face, he seems right at home, in his natural element.

In many ways he has had the quintessential naval officer's career of the latter half of the twentieth century, changing his path with the times and with the nature of world conflicts. His resume mentions his sea duty assignments to the destroyers USS *John R. Pierce* and USS *Damato,* among others, his move into the engineering duty officer community after finishing his M.S. in mechanical engineering, his postings to a variety of ever higher and more responsible positions in the naval shipbuilding program area.

What is missing, however, is far more subtle and in ways more interesting, certainly more in tune with the guy up on the bridge in the middle of the night staring at the water—and at the end of his naval career. Spend some time with him, and he may tell you what it was like in Vietnam in the early '70s, training Vietnamese naval academy graduates in amphibious assault operations for what even then everyone in his community knew was a war that was winding down. He might mention covert nightly intrusions into Laos and Cambodia, seeing off the SEAL teams. At the time I was there, he says in a typical burst of Navy doublespeak delivered with a smile, it was all classified and we weren't there doing that stuff. . . . I did a lot of unique operations that . . . it's not appropriate to talk about.

When I went into the Navy, he says, when I look back on it I don't think I was mature enough to recognize a lot of the things about serving the country, what freedom is all about. You don't recognize what it is until you have it taken from you. When I went to Vietnam, when I was in the Mediterranean, on North Sea deployment way away from home, thousands of miles at sea, the only thing is you and your ship and your crew. You get a bond out there that's really amazing, unique. You get very tied to your ship. You can talk to any sailor—officer or enlisted—and they *never* forget what ships they have been on. It's in their blood and they know it.

There was a time I remember, we had gone to the Western Pacific for nine and a half months on deployment. I had a wife and one child at home, and being separated from them was . . . not fun. I'm thinking, why me, Lord? Why am I out here in the middle of the Pacific Ocean away from my family. The Vietnam War was starting to wind down, they were going to sign the Paris Peace Accords, why me, Lord? I thought through that, and I was able to come up with an answer.

There are many more, Captain Staples continues, his eyes far gone into the memory, many many before me who have gone and done the same thing so that we can live in a free society, so we can enjoy the freedoms that we take for granted. Many died for those freedoms. So the sacrifice I was making was a limited one, really inconsequential in the bigger scheme of things. There isn't a time I sit down on a holiday—Thanksgiving or Christmas or Easter—I don't think there's some guy on a submarine or aircraft or a ship, an army platoon or deployed far away from his family. He's out there doing it, too. Just like I did and those before me.

Captain Staples pauses, perhaps reflecting on his own absences, the fact that he has moved his wife and three children twenty-one times in twenty-seven years all over the country. We typically, he says, are now doing six-month deployments. We just couldn't do it in the '70s. We had too many commitments and we didn't have a large enough force to do it, particularly with the Vietnam War. Some of the deployments were a year or longer. That's a long time. It's very tough on a lot of our people, when you come back and try to reestablish yourself within the family. An absence of six months to a year is a long time. You've lost something. It's hard to explain. You have to really work at it, and, at the same time be forced into a move to a different location. So you are trying to deal with the emotions of a move and a relocation and what that's doing to your family and trying to get the bond with them again.

He's also come to terms with more difficult, moral aspects of Navy life. He is, after all, the man on the spot responsible for shepherding every DDG-51 built in Bath on his watch from drawing board through launch and testing. And the DDGs are, after all, one of the most lethal ships in the surface fleet, carrying torpedoes, Tomahawk cruise and anti-submarine missiles, a five-inch deck gun, two 20-mm rapid-fire Phalanx cannons, and soon, two heavily armed Seahawk helicopters.

I think it's a *great* thing, the captain says almost defiantly of his role as a builder of warships. A noble job, he continues. The way I look at the ship, the ship is a protector of freedom, a repre-sentative of the nation and the influence we want to have on the world. If it can do its mission and never fire a weapon against another enemy target or kill anyone wherever, it's done its job. A lot of ships in the Cold War did exactly that. They were out there, day after day after day, in remote regions of the world on their mission ready at any time. And then they got decommissioned! To me, they were successful.

Eyes on the calm water ahead, he bemoans the fate of the Navy today. Reagan's much-ballyhooed six-hundred-ship Navy has gone from a strategic goal to a slender hope to an outright joke. The Navy had five hundred and forty commissioned ships in the mid-'80s, Captain Staples observes, now they have three hun-dred and forty, and the build rate can't even sustain three hun-dred. We're building six ships a year when we need to be building ten ships a year. His voice drops lower, and he turns his head to gaze out the window. It's been a great Navy career. I feel like it's time to move on. I'm not going to make admiral, and now I've had a command. And this is exactly and precisely the job I wanted for a command, it's amazing. The sadness will probably surface at the change of command, when I retire. I've done this all my adult life. This is all I really know. I love the Navy. It's a great service . . . but I'm looking forward to a second career.

11

COMMISSIONING

Perhaps the most important reason our Navy continues to be the greatest in the world [is] the ordinary sailor, whose willingness to do whatever is asked of him has never changed.

—*Commander James F. McCarthy, Jr., first captain of the* Donald Cook

It is often said that the United States Navy is the most traditional of the armed services, right down to their uniforms, whose basic design has changed little throughout the modern era. To the outsider, the rituals of daily life in the Navy—from the setting of the watch to the shriek of the bosun's pipe to the ringing bells by which time is kept on the ship—often seem impenetrable and antiquated, holdovers from a slower era when battles were fought at the pace of a horse or the speed of a sailing ship rather than within the nanosecond response times of the Aegis guided missile system.

Yet ceremony, ritual, tradition—these are some of the finer, more atavistic arrows in the Navy's quiver, tools that the Navy uses quite consciously in cementing the loyalty of crew to ship and captain, captain to service, service to country. Nowhere is this more evident than at a ship's commissioning, when for the first time the great gray leviathan is revealed for all to see as what it is—a warship and its crew, caught in a fleeting moment of perfection, beauty, almost innocence, before they set off to do battle with the sea and whatever adversaries their mission puts in their path.

The USS *Donald Cook,* December 4, 1998, at her commissioning in Philadelphia. After Mrs. Laurette Cook gives the order: "Officers and crew of the *Donald Cook,* man our ship and bring her alive!" the crew begins to stream up the gangway and come to attention along the rails.

On December 4, 1998, at Philadelphia's Penn's Landing, where the USS *Donald Cook* is to be commissioned, the setting, the weather, all are as if ordered from Central Casting. Penn's Landing forms a dramatic open-air amphitheatre, its interlocking, scalloped tiers of seating descending gently in a wide arc from busy city streets down to the swiftly flowing Delaware River.

Filling the eye from every seat in the house are the sleek, rakish lines of the USS *Donald Cook,* looking as if it is about to leap away from its moorings, and making even the most dunderheaded observer understand why destroyers are often called the greyhounds of the fleet. Bright, multicolored flags fly from stem to stern up over the very top of the radar mast, every rail is hung with red, white, and blue bunting. The ship glistens, nary a blemish of rust on its acres of gray paint, not a line uncoiled, every rare bit of brass polished to a golden sheen. The ship's outward weapons have shed their shrouds and display themselves almost defiantly, while here and there sailors in dress blues put chairs into place and guide distinguished visitors to their seats.

By early afternoon, with the sun high in a cloudless blue sky, the temperature has climbed to 70 degrees, and the audience is getting restless. The Navy band plays Sousa medleys while down on the quay the ship's crew begins to assemble in ranks. Officers, resplendent in white uniforms, flash gold from the braid of epaulets and on the bills of their hats. On the chests of the highest of the high brass—there are seven admirals present of various grades, one brigadier general, and a handful of captains—medals gleam above stomachs perhaps not as trim as they once were, while white-gloved hands reach out everywhere, clapping shoulders, embracing, shaking hands, touching cheeks. Here and there a ceremonial sword dangles, held carefully out of the way of the pressing throng.

Today will see not only the commissioning of this great ship, but also the start of all the celebratory events that will culminate in the greatest inter-service rivalry of all time, played out annually. In less than twenty-four hours, the Army/Navy football game will be held in Veterans' Stadium across town. Here every seat is filled, the crowd expectant and excited and still growing as the start of the ceremonies approaches.

Babies sleep oblivious in parents' arms, toddlers run amok up and down the stone steps. Grizzled old vets in service caps and with medals pinned to the front of rumpled sports jackets gather in small groups, exchanging war stories and looking out over the scene approvingly. Honoring the ship's namesake, the stark black-and-white POW-MIA badges abound, as do the caps of the Vietnam Veterans of America. Camera flashes blind the unwary, tears flow as mothers in their Sunday best behold their sons, magically transformed from gawky teens into proud sailors, and beaming fathers awkwardly embrace daughters dressed more formally than any prom gown could ever be.

Finally the ship's executive officer, or second-in-command, Lieutenant Commander Steven Lott, who will serve as the day's master of ceremonies, approaches the podium set up on deck amidships, and the audience quietens.

The DDG-51 class ships, Lott begins, are named for naval heroes and distinguished naval leaders . . . an honor roll of naval history, names like John Paul Jones, Alfred Thayer Mahan, and Stephen Decatur. We are proud to have Donald Cook take his place beside these distinguished figures . . . for his extraordinary leadership and courage while a prisoner of war in Vietnam from 1964 to 1967. His valor was memorialized with the award of the Medal of Honor. Our ship's motto, "Faith Without Fear," is inspired by his courage and faith in God and country.

Today's proceedings, he intones, are in keeping with a tradition that began with the commissioning of the Continental Navy's first warship. The first crew of the *Donald Cook*, formed in ranks on the pier, will always be known as "plankowners." In the days of wooden ships, it was customary for members of the commissioning crew to be presented with a plank from the ship's deck in recognition of their service. We don't have wooden planks to give away anymore, but the special connection between a warship and her first crew lives on . . .

Lieutenant Commander Lott reads the roll of honored guests, beginning with Cook's widow, Laurette, and his sister, Irene Coleman, then moving on to high-level representatives from the Navy's Aegis program and Surface Combatant offices who oversee the implementation of the system on every Aegis-class ship, from Naval Operations, from the Atlantic Fleet, from the Second Fleet, from Bath Iron Works and Lockheed Martin, the prime contractors, from the Supervisor of Shipbuilding in Bath. As Lott introduces them, they take their seats behind the podium.

In honor of today's principal guest and speaker, Admiral Pilling, vice-chief of Naval Operations, the lieutenant commander tells us, the Alpha Battery from the 109th Field Artillery will fire a salute. During the salute the guns will be loud, Lott says simply. Please do not be startled.

The drum roll begins, the band plays, and without warning the first of seventeen blank rounds crashes out from the gun battery hidden out of sight to the left of the audience, visibly startling nearly everyone. Each blast seems to vibrate the air, booming out, the wave of sound hitting the vast flat wall of ship's hull and washing back over the audience, dying away over the water, as huge clouds of white smoke begin to roll out over the assembled crowd. Seventeen times the field artillery fires, each round dying away through several seconds of aftershock and echo, during which it is all too easy for the observer to be touched, with the ship in front of him, by the merest imagining of war. Seventeen rounds and then silence.

Ship's company, atten-hut!

March on the Colors! Platform . . . salute!

The ship's flags, accompanied by an honor guard, are carried from the pier onto the ship to appear beside the podium, the band strikes up the national anthem, and Fire Controlman First Class Timothy Gilmore and Storekeeper Third Class Olimpia Herlo lead the crowd through a slow, sweet rendition of *The Star Spangled Banner*. Gilmore's voice is a confident, pleasant baritone, while Herlo's light soprano floats above. With a Southern church choir evidently somewhere in his past, Gilmore sings with a hint of gospel, which renders this so ubiquitous of anthems somehow different, unique to this time and place.

The chaplain leads a brief, hopeful prayer for ship and crew, and then Allan Cameron, BIW's president, strides up to the microphone, and his sturdy Scots brogue fills the air. What a great location, he begins, to celebrate the commissioning of this magnificent ship! Philadelphia is indeed an important place in the history of the United States. I am proud to be here today from another place of historical significance, the city of Bath, Maine, where for generations the men and women of Bath Iron Works have built fine ships for the United States Navy, earning the well-deserved accolade, "Bath-built is best built!" Cameron goes on to underline his company's unceasing efforts to modernize and improve, to change with the times to better serve Navy and country, then introduces one of the grand old men of the Aegis program, Rear Admiral George A. Huchting.

What a proud day this is! Admiral Huchting crows. What a great Navy day! It doesn't get any better than this as we look forward to the events of the afternoon.

The guests, caught up in the same magic that charges the admiral's words, clap their approval, with an unrestrained cheer here and there. The admiral cheerfully admits to being here under false colors, winning another laugh from the crowd, as he has just that day put in his papers for retirement, ending more than ten years with the Aegis program and a lifetime of service to the Navy.

Bath Iron Works, he goes on, builds the ship, but there are eighty-plus different programs that come together to make it a man-of-war. Those programs come together with parts and pieces provided from over forty states throughout our great United States. This, he gestures behind him at the ship, truly is a fabric of America. . . . It has been my pleasure to lead that team, and I regret that I am departing. . . .

Ladies and gentleman, he continues, his voice full of pride and emotion, I'd ask you to look at this crew for a moment. They don't come here by accident. They come from throughout our great United States. The Navy is a hiring business; we hire probably fifty thousand or so new sailors a year. We need the very best, the brightest people we can lay our hands on to join this profession. I leave this profession after thirty-six years; that's almost two full careers. I wouldn't change a day of it! I'm proud of it! I know that the mission our Navy does for you, the American people, and for the world is vitally important. . . . President Kennedy said several decades ago that if a man could say of his life that he had served in the Navy, that would be quite an accomplishment.

After a moment of silence during which the audience can contemplate together the absolute truth of that statement, Admiral Huchting introduces Admiral Henry Giffin, who is, his colleague asserts fiercely, a cruiser-destroyerman, a sailor's sailor, a seaman without equal, responsible for all the cruisers and destroyers in the Atlantic Fleet, and more important, a great old sea dog!

Doesn't it make you proud, Admiral Giffin begins, to be an American, to see these young sailors and this fine warship in front of you! Giffin, presently the commander of the Naval Surface Force of the U.S. Atlantic Fleet, hails today's guest of honor, the Vice-Chief of Naval Operations Admiral Pilling, as a kindred warrior in the great tradition of Navy destroyermen. The DDG-51 class of ships we celebrate today, Giffin tells the audience, are in large part a result of his farsighted work over the years. It is my honor and privilege to introduce the thirtieth Vice-Chief of Naval Operations, Admiral Pilling.

Just a few blocks away in the heart of the old city, Admiral Pilling begins, Independence Hall and the Liberty Bell stand as enduring symbols of the greatness of America, while afloat before us is a new and powerful guardian of America. And just like our national heritage being close by, the Aegis radar system, which is the heart of this magnificent ship, was developed about fifteen miles from here in an RCA plant twenty-seven years ago this month.

Donald Cook, Pilling says, is the newest ship in the longest-running surface shipbuilding program in the history of our nation. This guided missile destroyer is truly an awesome sight: 9,000 tons of warship, the most advanced electronic and sensor systems, the full array of armament that allows her to dominate the sea while projecting naval power several hundred miles inland. . . . She is a marvel of American industrial ingenuity. . . .

Pilling pauses, seems to look down on the ship's crew gathered on the pier below. But this wondrous technology, he continues, is only part of the story. Accomplishing the vital mission of protecting America and protecting our worldwide interests requires the highest caliber of young people our nation has to offer. Over two centuries ago, Captain John Paul Jones called upon the young, the brave, the strong, the free, to join him on the high seas. We are fortunate that his words remain alive and well in modern America, as proven by the remarkable men and women who stand before us as the commissioning crew and plankholders of the USS *Donald Cook*. They have worked tirelessly to make this ship and themselves ready . . . for the challenges that await them as they sail the world's oceans.

Admiral Pilling turns to the seated dignitaries behind him and looks the ship's first captain in the eye. Commander McCarthy, he says, from this day forward you will be known as "Captain," the most revered title in the Naval service. My compliments to you, Captain, and your crew. In appearance and performance you are all shining examples of what makes our armed forces and our Navy the best in the world! Pilling goes on to invoke Donald Cook's memory and honor, reminding the gathered that safeguarding America and upholding our highest ideals demands sacrifices on the part of our young men and women in uniform.

There is another American hero here today, the admiral finishes, who I must mention as well, someone who has also displayed courage, compassion, and an indomitable spirit. In those years of uncertainty following Don's capture, Laurette unselfishly devoted herself to providing comfort and hope to the families of the other prisoners and those missing in action while raising four children as a single mother. Laurette, I'm sure I speak for the officers and crew, the spouses and families, and every American who is inspired by the memory of Donald Cook, in saying that we are truly honored to have you as the ship's sponsor. The crowd bursts into applause, rising at first hesitantly, then as a whole, to offer this courageous woman a standing ovation. Finally, Pilling says as the last clap dies away, in light of tomorrow's game, I'm sure General Shelton would join with me to conclude with—beat Army!

Ship's company, atten-hut! Lott orders.

As authorized by the secretary of the Navy and for the president of the United States, Admiral Pilling intones, I hereby place United States Ship *Donald Cook* in commission. May God bless and guide this warship and all who sail in her. His words are followed by applause as the next order rings out.

Hoist the colors and the commissioning pennant! Signalman First Class Hunt, hoist the colors and the commissioning pennant!

Aye-aye, sir! All eyes turn up as the drums roll and two flags make their way from deck to topmost mast.

Captain, Lott reports, the colors have been hoisted and the commissioning pennant is at the main trunk.

Very well, Captain McCarthy acknowledges. He ceremoniously opens the envelope holding his orders and reads them aloud, then turns to Lieutenant Commander Lott. Executive Officer, he calls out authoritatively, set the first watch!

Aye-aye, sir, Lott responds. Officer of the Deck! Set the watch! Aye-aye, sir! comes the response. The officer of the deck, Lott explains, is the commanding officer's direct representative. While on watch, he is responsible for the safety and smooth operation of the ship and crew. The long glass [a brassbound telescope] is the traditional symbol of the officer of the deck's authority in a ship of the line. We are honored to have Donald Cook's sister, Mrs. Irene Coleman, here with us today to pass the long glass to our first officer of the deck, Chief Petty Officer Charles Rockwell of Houlton, Maine.

As he finishes, and Mrs. Coleman passes the shining long glass to the officer of the deck, the bosun's pipe sounds its two eerie notes, then trails away to nothing. Sir, the watch is set, CPO Rockwell reports. Very well! Commander McCarthy turns back to the microphone and ushers Laurette Cook to the podium.

Mrs. Cook, a diminutive, gray-haired woman with strong features and a ready smile, brings a very personal note to the ceremony, relating what kind of man her husband was, how deeply he loved his family and country. The assembled families and friends listen intently, some with bowed heads, as she tells in a few brief and simple words the story of a man, a father, husband, son, brother, and Marine who gave his life so that others would not have to.

Shortly after arriving in Vietnam, Laurette begins, Don volunteered to go on a dangerous rescue mission. A helicopter had been shot down. Don went with a group of Vietnamese Marines to aid the crew. The rescue party was ambushed, and Don was shot and captured. He became a prisoner of war on New Year's Eve, 1964. My family and I waited until the end of the war in 1972 to learn of Don's fate. The day the POWs were released, we learned that Don had died several years before. One of the hardest days in my children's lives was the day we watched the released POWs arriving at Andrews Air Force Base. My children—Karen, Tori, Tom, and Joseph—watched, hopefully, for the image of their dad, for him to come off the plane. But of course he never did.

She reads from a letter smuggled out of the POW camp where he was held, sometimes in a ten-foot-square cage, for the three years leading up to his death. "If I can sum up my thoughts I would like to leave this with you—Joseph, Karen, Tori, and Tom, especially them. Do what is right no matter what the personal cost. Love of God and man above all else. Don't judge yourself by others, but others by yourself. Life to me is so simple: there is life, death, and eternity. If we can't save our souls, what good is anything? This guides all my actions . . ."

To the commander and crew of the USS *Donald Cook*,

Laurette says after a pause, remember the man behind this great ship. As your motto states, have faith without fear. Godspeed!

The crowd begins to applaud, rising once again to honor this most amazing woman. At last, when all is quiet, Laurette looks down on the assembled crew, and with a wide smile on her face, calls out the order:

Officers and crew of the Donald Cook, man our ship and bring her alive!

Aye-aye, ma'am! the two-hundred-strong ship's company shouts enthusiastically. The Navy band breaks into a spirited "Anchors Aweigh," as the orderly ranks on the pier break up into two long files of sailors, snaking slowly up gangways at the bow and stern until they line the ship's every rail at spaced intervals, standing stiffly at attention. As the last of them take their places, the band segues into the Marine Corps Hymn: "From the halls of Montezuma . . ."

At an invisible signal the ship's sirens sound, its whistle blows, and Torpedoman's Mate First Class Valarie Williams appears beside the aft torpedo launcher with its three stacked tubes. Valarie swivels them out to face the crowd, and fires slugs of compressed air out of each in turn, releasing compact loads of confetti that sail, in a clump, twelve feet out over the audience and descend in great gobs onto the same small group of unsuspecting people.

Honored guests, ladies and gentleman, Lott says, the crew of USS *Donald Cook* salutes you! The sailors behind the rails come to attention as one, saluting smartly, their faces by turns proud, joyful, solemn, relieved.

We are proud to serve in your great Navy, Lott continues after a pause to enjoy the moment. Captain, the ship is manned and ready!

Captain McCarthy turns to his immediate superior, Captain M. J. Miller, who commands the destroyer squadron that the *Donald Cook* will soon join, and salutes.

Commander Miller, McCarthy says formally, I have assumed command of the ship. The ship is ready, and I report for duty at Destroyer Squadron Twenty-two! He turns to the crowd and begins to speak in a low and pleasant voice, welcoming families and honored guests. He hesitates, seems to take in everyone and everything around him in a sweeping glance, then continues on a different, more reflective note.

On a hot July day two hundred and twenty-two years ago, Commander McCarthy begins, fifty-seven men met here in Philadelphia. Their goal was to shed the yoke of tyranny and gain liberty for their families and their futures. It has been our duty . . . to carry on that tradition of freedom, and to insure its blessings are passed to each successive generation.

As the captain of the *Donald Cook* continues, it is clear that he speaks as one with the most profound respect for his ship and its namesake, for its crew and the institution they together represent, the awesome technology, he tells us, that will help sustain the backbone of our Navy's fighting force, a salient example of American resolve in our own defense.

McCarthy reminds the gathering of the importance of honoring not only the ship and crew, but the memory of the man for whom it is named. It is important for us to do this, he tells us, [or] we will suffer the fate warned of by President Calvin Coolidge: *A nation which forgets its heroes, will itself soon be forgotten.* Colonel Donald Cook served his country when he was called upon and, ultimately, gave his life that others might be free. . . .

The USS *Donald Cook* brings to the fleet, he continues, all the marvels of modern technology. The 9,000 tons of American fighting steel you see before you on this pier . . . represents the best of American ingenuity, perseverance, and determination. There is, however, an inherent danger in relying too heavily on our technology. Time and again history has shown us . . . [that] any gain or advantage that we have today is at best transient and will require continued evolution to insure our futures.

There is one exception to this reality, he concludes, and it is perhaps the most important reason our Navy continues to be the greatest in the world. We have one weapon we have had since John Paul Jones met the British frigate, *Seraphis*, in 1779. No one has ever come up with a counter to . . . the ordinary sailor, whose willingness to do whatever is asked of him has never changed. Our Navy is comprised of four hundred thousand men and women who represent the finest this nation has to offer. The group assembled in front of you today is, in my estimation, the three hundred best of those elite Americans. They're smarter, more dedicated, and they're harder-working than we have a right to deserve, and it's their contribution as well as the sacrifices that are made by their families, that make our Navy ultimately invincible.

To an audience comprising several thousand family members, mothers and fathers, sisters and brothers, grandfathers and grandmothers, these last words hit home. The ship will make its way south in a few days, heading for Puerto Rico and several arduous months of weapons and firefighting qualifications before joining the Atlantic Fleet for an initial six-month mission. The brutal reality of a sea posting means that it may be nine months to a year before fathers see their wives and children again, before mothers are reunited with their young children.

Commander McCarthy finishes with a more celebratory act, meritoriously promoting four sailors on the spot, outside the normal channels, by the power vested him in the Navy as the captain of the *Donald Cook*.

The Navy band plays a final march, the crowds disperse, some to tour the ship, others to the nearby Convention Center for an afternoon reception. Sailors drift off the ship, seabags slung over shoulders, Dixie cups (the round, flat white hats with the upturned brim that are part of formal dress) at jaunty angles, to spend a last night with parents, lovers, friends. The vast tiers of seats are suddenly empty, here and there a forlorn young couple, the young man or woman in uniform, locked in what seems, after the emotion of the ceremonies, a desperate embrace. Programs litter the paving stones, stirred by the gentle breeze. Down by the ship, camera crews coil cable and photographers pack their bulky bags. The band, so bright and cheerful, has vanished, taking the music with it. One rite of passage for the ship and its crew is over, and what lies ahead is still, thankfully perhaps, unknown. There is no question, however, that these 320-odd men and women have, in every effort leading up to this moment, accomplished that impossible miracle that must happen in order for every ship to fulfill its destiny—they have, in the captain's words, breathed life into what would otherwise be just 9,000 tons of metal, glass, and plastic.

EPILOGUE

This ain't a cream puff factory, it's a shipyard.

—*Jay Bailey, VP Production, Bath Iron Works*

The first time I saw Bath Iron Works launch a ship into the Kennebec River, I happened to be driving over the Carlton Bridge just up the river from the yard and noticed the colorful flags and the crowds below. An hour later I stood with a small group under one of the piers of the bridge as the USS *Sullivans*, an Aegis guided missile destroyer, slid into view from behind its staging and splashed into the river. It was the largest moving man-made object I had ever seen up close, a haze-gray monster, and it was heading broadside right for us.

The ship drifted closer and closer until, wherever the gaze fell, there was a flat wall of steel. We looked at one another without asking the obvious question. Finally, someone said it was coming awful close to the bridge, and everyone laughed uncomfortably. People on deck waved at us, the tugs surrounding it belched more black smoke, and the ship slowly, miraculously began to turn, then to move off downriver. I left with the overwhelming desire to get closer to this amazing thing.

Nearly a year later I stood anxiously at the main gate of the yard, waiting for Tim Vear, an electrician who had agreed to show me around for a few hours one afternoon. Tim arrived, a short, squarely built man in a green hardhat, which I, not knowing that the color signifies the wearer's trade, took to be another sign of his cheerful demeanor.

Three steps into the yard and my heart was beating faster. The sights and sounds of a shipyard at work washed over me, and Tim laughed at my swiveling head and darting eyes. You learn to pay attention to what is dangerous, he shouted into my ear, pointing overhead to a huge load suspended from the twenty-five-story arm of a crane, then pulling me out of the roadway as a sixty-four-wheeled transporter laboring under 100 tons of completed deck-house passed within a foot of us. Standing in the open doorway of an assembly building, I noticed a welder gesticulating at us wildly from twenty feet away. He was using a gouger with a copper-sheathed welding stick in its teeth to remove a steel clip that had been welded to the deckplate. Running high-voltage electricity through the stick creates a high-temperature arc that melts the old weld, at which point the welder adds high-pressure air to blow the molten metal out and away. Tim steered me clear as the welder's hood went down and his gouger hit steel, bouncing sparks of molten steel off the spot we had just occupied.

That first day, I didn't know where to look, what to watch out for, where to turn next. My overwhelming impression was one of chaos, of grime-covered men in coveralls doing very dangerous things with old equipment amid piles of junk and rusting steel. Tim confirmed that people do get hurt here all the time. Very occasionally, they die. In one building, the cadmium white of a plasma burner flared as a bulkhead took shape in front of us. No matter how I looked at it, I couldn't figure out what it was—my first glimpse of an upside-down unit! A gray-haired workman, eavesdropping, told me that you can always tell the new guys on the job because they can still tell left from right. The old-timers are so used to working upside down that they get them backwards.

And so I began to spend more time at the yard with mechanics from various trades, and very quickly discovered, as a tin-knocker told me one day, that you don't know squat until you've worked there ten years. What I *had* figured out was that the mechanics couldn't care less who's running the company, what conglomerate just bought it, or what the Navy thinks; they love the ships and their making and launching as fiercely as they love their kids. Don't tell management this, Tim Vear had confessed to me that first day, but I'd work here for free. I hadn't believed him then, but now, after hanging around with these guys, *I* was ready to ask about job openings.

The yard has that effect, drawing you into its mysteries little by little until you are consumed. As I stood beside a ship about to be launched one day, a shipfitter laid his hand on the hull like he would on his boy's shoulder at a baseball game. Building and launching this ship, he said, it's like having a big secret which no one else knows. *You* know, he said, because you work here every day, ten degrees or a hundred degrees, three feet of snow or a week of rain, you gotta finish the job. But they—he gestured to the empty seats where the spectators would soon sit—you see their mouths drop open and you think, they'll never know. I looked at him, this burly guy with a walrus mustache and a Harley-Davidson emblem on his hardhat, with dirt under his fingernails and small white scars from hot sparks dotting his hands and forearms. I wanted to know his secret. Badly.

My second launch, six months later, I arrived at the yard in the middle of the night for a wedge-driving, when nearly two hundred men with long iron rams drive oak wedges under the wooden cradle, driving it up to embrace the ship for their short trip together into the water. I was looking at history, at something the mechanics' grandfathers and great-grandfathers had done all the way back to the first ships built on this river so long ago. Those few hours next to the river before dawn, as the rams crashed against the wedges and set the hull to booming and the yard whistle sang out the rallies, I felt very deeply how much these men had tied up in this ship—a trade and a living to be sure, but also an identity, a piece of their hearts. With this particular ritual soon to disappear, and the yard itself a David struggling against forces it couldn't begin to control—global competition, Senate infighting, the New York Stock Exchange—the urgency of capturing these moments intensified.

The first Saturday of October at 3:22 P.M. I was standing far down the launching ways, almost to the water, right next to the trigger arm, a crude metal pipe whose business end had been slapped with red paint. The ship towered over me, blotting out the weak sun. The concrete building ways running beside the ship from bow to stern had been just hours before cluttered with men, machinery, and most important, with the angled shores then supporting more than ten million pounds of steel ship. Now the concrete was empty. The destroyer rose up from the ways, its belly securely resting on the wooden cradle on which it would slide into the water, its brass propellers gleaming dully at the stern, which would enter the river first.

The Launch Master, Dean Atkinson, stood on one side of me. Chief Hull Engineer Erik Hansen hovered on the other. The dignitary who would actually throw the lever took up her position, careful not to touch that slender handle on which 300 tons of pressure had built up over the last hours.

At 3:26 Dean's radio crackled. *Secure for launch. Platform to trigger: Launch. Aye. Launch!* Dean answered.

The dignitary touched the trigger. It flew out of her hand and crashed into the concrete. I thought something had exploded. Rick Libby bodily dragged the sponsor away from the ship, which had begun to move at first slowly, then very very fast. The cradle boomed and popped, groaning under its great weight and, now, speed. Six feet from my face, five hundred feet of hull raced by at 20 miles an hour, grease ribboning out from under the lip of the passing timbers. My brain said, Move! but my back was already against the staging. The bow passed us, and we all scrambled out onto the ways. No one said a word until the bow dipped suddenly. It had run out of ways. The ship floated free for the first time.

Walking out to my car that afternoon, I realized that I was shivering from the cold, and wet through with drizzle. My face burned from too much hazy sun, and my stomach was cramping from having gone from 4 A.M. that morning without food. I just hadn't noticed earlier. I had been seduced, utterly, by the yard, by the men working as if driven by the devil, by the ship itself, a cold thing of steel coming to life before my eyes. I had gotten a long, privileged look at the arcane art of launching a ship, and I had learned enough of the shipfitter's secret to know that I hadn't the right to more without actually taking a job there.

What I could do, however, was to tell this story, the story of a place with a past and a *present*, a bright shining beacon to show, in these cynical times, what men and women are capable of when history and tradition are respected, when these noble concepts usually so casually bandied about not only live on in daily life but are necessary to it, embedded in the sunrise-to-sunset workaday tasks of building a ship. The bottom line at Bath Iron Works is, still, that this warship, however crammed with this morning's technological innovations, cannot come to life without the hard labor of men and the centuries of cumulative knowledge that is the craft of shipbuilding. Is it so surprising, then, that their hearts get so caught up in it, too? This particular intertwining of the work of the hand and the work of the heart, in our world, is unusual and worthy of our attention and respect.

Some might think that this is just a glib romanticization of what is, at the end of the day, an antiquated shipyard that has survived more out of luck than anything else, a place where the work is dirty and dangerous because they can't afford "modern" tools and haven't the necessary corporate commitment (or capital) to modernize their methods. None of that is true.

Five years ago you would have said BIW was on its way out of business, says Allan Cameron, BIW's president and architect of the $210 million modernization campaign that he convinced General Dynamics, the shipyard's corporate parent, to fund in order to recapitalize the shipyard and move it into the next century. Cameron, blue-eyed and ruddy-cheeked under a thatch of wiry gray hair, has a presence larger than his moderate stature, an infectious energy that surrounds him and draws in those around him. His speech, a rich Scots brogue, betrays his origins on the River Clyde in Scotland, a place much like Bath, with a long and storied shipbuilding tradition. Cameron speaks with the plainness of a man who began work as an apprentice shipfitter at age sixteen; a man who, during thirty years of shipyard experience, has seen his industry shrink from, in the U.S. alone, nearly thirty shipyards to a bare handful today. He is, in short, a passionate shipbuilder with both deckplate and management experience, regarded by many as a perfect candidate to lead the company into the future.

It's not too long ago, Cameron says thoughtfully, that we were declaring to all of our employees that, if we couldn't turn things around, we would be out of the DDG destroyer program because we were no longer an affordable shipyard and we could no longer deliver a ship on time. Today, we're the lead yard for the DDG, we're part of another team for the LPD–17 Marine amphibious assault craft, we're part of the lead team on the 21st Century Surface Combatant. We are one of the most flexible, the most capable shipyards in the country for other major giants like Lockheed Martin and Raytheon, who want to team with us on new and exciting platforms for the future. That tells you we're doing something right.

Cameron has a very clear set of goals, a vision, and is willing to go to the wall to bring that vision to life. During the hard-fought contract negotiations of August 1997, the situation was not good at all. Union President Brian Bryant and his negotiating committee were actively encouraging a strike; and the Not-So-Silent Protest, workers taking hammer to steel five minutes every hour on the hour, was disrupting work at the yard—and the normal life of the town. This came at a time when Cameron was, in order to help finance the modernization, lobbying the town of Bath for $81 million in tax deferments, the state for about $100 million in tax relief, and the Navy for the good-faith release of monies held back as contract guarantees. Finally the approximately five thousand Local 6 members, the men who build the ships, voted 51 to 49 in favor of management's proposal. While the offer did include some improved health and pension benefits and $1,500 worth of bonuses, it had a real rise in wages of only 25 cents an hour for the three-year life of the contract.

Talk to Cameron for any length of time, and you will soon conclude that this man is no dewy-eyed romantic. BIW has one customer and that is the Navy. It is a demanding customer, one with, in his words, a voracious appetite for making changes. He looks at structural changes and reforms in the Navy's acquisition process and sees a business opportunity; namely, that BIW and its industry partners are now being asked to perform (and get paid for) functions previously carried out by slow, expensive, and cumbersome government bureaucracies.

Acquisition reform, he says, is stopping the traditional long-term process. Instead the Navy talks to industry, says, *you* develop it from day one, *you* get me that state-of-the-art technology, and *you* get me leveraged commercial off-the-shelf equipment and practices and processes [versus more costly traditional military specification procurement practices], and give me a war-fighting ship much sooner than we could under the traditional process. And what's driving that? Affordability. The government can't afford the naval sea systems command infrastructure, the number of military bases, those time-honored institutions and organizations that are prevalent throughout the Department of Defense.

Taking on the concepts design and engineering for proposed new platforms—new ships and all of their systems—is what Cameron calls growing the business on the front end. Another part of the business, Cameron says, is that our Navy customer is asking us to become the life-cycle support agent to maintain those ships and to upgrade them for their thirty-five-year life. Traditionally, the shipyards didn't do that. We were the planning yard, and all we did was plan. Now it's maintenance, it's operations, it's upgrades, it's technology insertion.

So that is Cameron's vision. Not necessarily building more ships, and certainly not building commercial ships, but doing more of the work on each class of ship, both up front in its design and long-term throughout its life. That's our focus, Cameron concludes. Do that, and we can be a tremendously successful company for a long, long time.

Hand in hand with the conceptual revolution at BIW is the very real change out on the deckplate, in the way the shipyard builds and launchs ships. The stresses of this are already evident, both in larger conflicts with the union over the reorganization of the work itself, and on a more personal level, out on the deckplate, in the outlook of the mechanics and the lower-level supervisors. With pride in craftsmanship and an unmatched work ethic as central to the company's competitiveness as the modernization, BIW faces the question of how to keep the spirit of tradition alive through the next generations of shipbuilders while adapting to new methods and new times.

Everywhere I went throughout the shipyard, I sought out those mechanics who knew their craft, who took pride in it, and who could articulate both the intricacies of their work and its meaning to them. All of these workers, invariably the most gifted and respected on their crews, most with fifteen if not twenty years on the job, had been offered the "white hat"—a supervisor's position. All had refused.

These very capable "elders" (mostly in their late thirties or early forties), the very workers who one would imagine the company wanting to move up the ranks, told me that being a supervisor, especially in the middle ranks and above, was just not worth the hassle, the stress, the loss of respect in the eyes of their comrades. Not only would they be giving up the protections of their union seniority, they said, but the company would also not support them, either with enough training to make them effective managers or when push came to shove with the union over personnel or work issues out on the deckplate. Too often they had seen one of their own who took the bait end up abruptly reassigned to another job after a conflict with the union, or put in charge of an area in which he had no experience. I ruined my back and my hands working here twenty year, one fitter told me, so now I'm gonna go for the ulcers, maybe a heart attack, if I put on the white hat?

When I began writing this book in 1997, BIW's future was not at all certain. Today the company is preparing to sally forth into the next century with new facilities, new equipment, and fifteen years' worth of future work already in its pocket. They will build at least one new kind of ship, the LPD amphibious assault ship, in consortium with Avondale Industries and Hughes Electronics, and are still leading the pack in the competition to participate significantly in the Surface Combatant 21st Century program, the major shipbuilding program of the twenty-first century. They are led by a visionary president with a passionate love for shipbuilding combined with a hard business sense and an iron will. What remains to be seen is whether or not the white hats can bring the mechanics along with them. I find it hard to believe they won't, as they have for more than a century, muddle their way along to a solution somehow. Shipbuilding, I was told once, just runs in the blood around here.

AUTHOR'S NOTE

In writing this book I used several common nonfiction journalistic conventions that the reader should know about. First, for clarity and brevity, I collapsed what may have been up to a month of interview and observation of a single mechanic or small group of mechanics into what appears here as a scene or a part of a chapter. Everything you see, I saw. Everything you hear, I heard. There is no invented or imagined dialogue, nothing made of whole cloth.

When possible, I verified controversial stories and facts with at least two independent sources. When accounts conflicted, I included as many differing points of view as was possible.

Most readers will realize right off that I have dispensed with quotation marks, instead letting conversational markers of the "he said," "she concluded" variety set off direct speech. Because of the need to impart so much information in the midst of any given speaker's words, the page would have been unreadable chicken scratch with all of the paragraph jumps had I used normal punctuation. I thank Sebastian Junger for indirectly steering me in the direction of this unusual narrative style with his effective use of it in *The Perfect Storm.*

Finally, I beg the indulgence of shipbuilders everywhere, but especially at BIW, who will tell anyone with an ear that no book can show all that goes into such wonderous creations as are ships. I have omitted large amounts of technical detail and sketched briefly where others might have lingered, for the sake of brevity and to keep the narrative interesting.

ACKNOWLEDGMENTS

Taking on a subject of this complexity would not have been possible without the help of many, many people at all of the Bath Iron Works facilities, some in white hats and most not, some in offices, but most in drafty workshops or out on the deckplate. I am indebted to many who wear the uniform of the U.S. Navy or are associated with the Navy through both the Bath Supervisor of Shipbuilding, Conversion, and Repair office and Pre-Commissioning Units across Washington Street from the yard. My thanks to all of you, and my apologies in advance for any misapprehensions and mistakes, which are wholly mine. My only regret is that, in a book of this size and kind, I could not include more of you and your stories.

I must first doff my hat to Julie Phillips, former manager of BIW's Public Affairs office, who opened the door for me and then got out of the way, interceding at various crucial points to make my long road a much easier one to travel. Thanks to those others so crucial in getting this off the ground both at the yard and as a proposal: Ed Moll, Larry Albee, Brian Bryant, and Blake Hendrickson.

I owe a major debt of gratitude to my "spirit guides" in key areas. Tim Vear and Troy Osgood first took me in hand and gave me the overview. At the Hardings plant Don Lamson and Dave "Cookie" Cookson began my education in plates and shapes while Jim Alexander and his Bending Floor crew, especially Peter Marshall and Kingsley Barnes, showed me the intricacies of bending and straightening steel. Ben Favreau, Larry Caton, Don Hyde, Charlie DuBarr, Joe Tetreault, Malcolm Springer, Bert

Wyman, Kenny Garland, Virginia White, Fred Hannah, Mac McKeown, Mike Atwater, Michael Proshinsky, Reggie Beaulieu, and Larry Carlon tutored me in everything from "taking a heat" to paint application to how to bend aluminum. Harold Hinckley, Pat Roderique, Dave Bisson, Tim Hudson, and particularly Cal Sutter showed me how to weld; while, at the yard, in the Welding Technology area, Rick Marco, Dave Forrest, and Clyde Hinckley taught me welding's whys and wherefores.

To help me better understand the ship design and construction process, Russ Hoffman and Craig Whitman both spent hours of their valuable time, for which I am grateful. This stood me in good stead when I moved to the ways, where Erik Hansen, Dean Atkinson, Tom Niles, and Chris Medeiros passed on a trove of information about cradles and launching, took me everywhere I asked to go, and let me tag along from start to finish on the four launches that took place during the writing of this book. I thank them for their stories and good humor, and will not soon forget the thrill of dawn, launch day, down by the river. David Bosse, Dan Athearn, and Steve Richard, all pipefitters, and "Lippy" Lippoth, too, contributed to making the launch chapters real while Rick Libby, Director of the Ways, also aided me here.

Gary Blackler, a truly knowledgeable and dedicated individual, guided my way through the labyrinthine workings of the Assembly Building, Pre-Outfit 2, and Five Skids. Charlie Barnes and his second-shift crew took me in hand in AB, with Mike Matthews and Randy Fisher teaching me as much as I could absorb about shipfitting and unit assembly. As well as to Gary, Charlie, Randy, and Mike, I owe thanks to Jason Cloutier, Merle Witas, Eric Bowman, Matt LaRochelle, Luke Arsenault, John "Gummy Bear" Collins, Ted Kramer, and Gary Havlicheck. Thanks to all the riggers inside and out who shared their vast expertise on moving large objects, especially to Sherman Mitchell, Chuck Wright, Pat Toth, Mike Knight, Lou Murray, Terrence Flaherty, and the anonymous second-shift Crane #15 operator. Buffy Knight, Richard Douglas, Mike Shaffer, Burt Boisuert, Pete Watson, David Huggins—all contributed more than they know to Chapter 6.

Jay Bailey, VP Production at BIW, understood enough of what

I was doing to permit me to go to sea on the *Donald Cook*. At sea, my guide was Captain Earl Walker, whom I'll sail with any day. Mark Colby, BIW's trial coordinator, and Wally Pitcher, Joel Culver, Rick Townsend, Joe Pinette, Mike Zubiate, Dave Ward, and Mike Smith also advanced my knowledge of how a ship is tested and run. Thanks to Tim May, Jim Balmer, Al Reid, Bert Gilliam, and Tom Gregory for the 2 A.M. engine room and CCS tour, and to the anonymous Navy INSURV personnel who answered my questions and explained their jobs.

This book would have stopped at the pierside without the help of Lieutenant Commander Mike Anderson, who taught me so much about his ship, while Commander James McCarthy shared his years of Navy experience. All of my Navy encounters were orchestrated by the very able and overworked PAO, Ordnance Officer David Ausiello. FC-1 Tim Gilmore laid out the Aegis system while QMSR James Sippel, TM-1 Valarie Williams, ST-3 Lori Naron, and ET-1 Beth Holz gave me a more personal, valuable insight into their world. SUPSHIP's Captain Ralph Staples shared a career's worth of knowledge, and his assistant, Barbara Lagassé, helped with documentation. Bear Hastings at PreComm was also a big help.

Onboard the USS *Donald Cook*, I also had help in one way or another from the following: BM-2 Edwin Bennett, Senior Chief Oakley, YN-2 Torase Williams, MR-2 David Lanham, FC-1 Vilano, First Lieutenant Jennifer Friend, Lieutenant J. B. Mustin, Lieutenant Sonya Perry, SM-2 Tiffany Chewning, QM-1 Timothy Rochford, Lieutenant James Katin, QMSN Marvin Stroud, GM-3 John Morris, CTR-1 Travis Henson, FC-3 Matthew Smith, BM-2 Allen Lovelace, and MSSR Andre Christian.

The town of Bath has three hard-working, capable representatives in the town manager, John Bubier, assessor Paul Mateosian, and planner Jim Upham. They helped me understand the town-gown aspect of life in Bath just as the Bell-Hoerth's—Mary Ellen and Tom and their two children Emily and Carrie—told me of the more intimate side of life there. I thank them for their willingness and enthusiasm, shining examples all of the promise and energy of this small town on a river.

BIW's president, Allan Cameron, and his VP Human

Resources and Public Affairs, Kevin Gildart, opened my eyes to some of the larger, corporate issues defining BIW's present and future, and I thank them for a view I otherwise would not have had. Sue Pierter, Director of Communications, came through in the last months with invaluable help, as did Rusty Robertson in her office and Tom Travers at CSC.

I must also publicly pat the back of my agent, Jane Chelius, not only for the superb negotiating which turned a good contract into a great one, but for her willingness to do battle in her client's name no matter how small the issue. On a related note, I must thank John Faulkner and Sylvie Wyler, in the basement of whose gallery the proposal for this book was written, and Joel Knee, secretary of Transportation, and Wanda Klamanski, chief transcriptionist, for their efforts. My co-workers at Bookland of Brunswick encouraged me at the bleakest, as did Stuart Gersen and Kathy Hayden at the Mother Ship in Portland. Stewart MacLehose at the Patten Free Library filled in historical blanks and steered me to others in the know.

Anthony Walton presented me with the gift of the title, which truly has a value beyond measure, and helped define for me what success as a writer is and can be through his fine example.

I am lucky to have the world's most supportive family, and I thank them all for the faith they showed through times thick and thin, and for putting up with the absences and distractions.

INDEX

BATH IRONWORKS SHIPYARD   Bath, Maine

KENNEBEC RIVER